THE TELECONFERENCING HANDBOOK

A Guide to Cost-Effective Communication

by

Ellen A. Lazer
Martin C.J. Elton
James W. Johnson

with

Al Bond
David Boomstein
Eugene Marlow
Robert D. Rathbun
Bonnie Siverd

Knowledge Industry Publications, Inc.
White Plains, NY and London

Communications Library

The Teleconferencing Handbook: A Guide to
Cost-Effective Communication

Library of Congress Cataloging in Publication Data
Main entry under title:

The Teleconferencing handbook.

   (Communications library series)
   Bibliography: p.
   Includes index.
   1. Teleconferencing.   I. Lazer, Ellen A.   II. Series.
TK5102.5.T425   1983   384   82-18739
ISBN 0-86729-022-6

Printed in the United States of America

10   9   8   7   6   5   4   3   2   1

Table of Contents

List of Tables and Figures

List of Illustrations

ACKNOWLEDGEMENTS

I am especially grateful to Dr. Martin C.J. Elton of the Tisch School of the Arts, New York University, and James W. Johnson, president of TeleConcepts in Communications, Inc., for their helpful reviews and criticisms of the draft chapters of this book. Karen Sirabian, associate editor, Knowledge Industry Publications, Inc., did much to make this book a cohesive and handsome publication. Finally, thanks are due to the many industry experts who answered questions and provided information, and made available photographs and illustrations for this book.

Ellen A. Lazer
January 1983

Introduction

Teleconferences—electronic meetings—are in the forefront of today's new communications technologies. They are part of the Information Age, the Electronic Age, the Office of the Future, the Television Age, the Communications Revolution, the Computer Age. However one chooses to categorize teleconferences, they are already standard features in many of the largest corporations and institutions, and are fast becoming communications tools to be reckoned with by all kinds of American organizations: businesses, universities, unions, medical centers, government agencies, professional associations, social service departments, even the State of Alaska.

Yet the number of people who have actually experienced a teleconference—particularly a video conference—is still small, and the number who have actually organized or produced one is even tinier. Making the transition from an operator-assisted conference call or a closed circuit TV program to the logistics of an audio or video conference is not an easy task. Just deciding upon a teleconference—which kind of teleconference—requires a lot of knowledge and judgment.

This book is written for the many individuals who are likely to be involved in audio or video conferencing—those who may decide that it's the best way for them to train, market or meet some other informational communications need. The manager who decides that a teleconference is the best tool for a particular communications need must then know how to organize one. Therefore we provide detailed but nontechnical explanations of the major questions that a prospective teleconference user needs to answer:

- What is a teleconference?
- How does it work?
- Which kind of teleconference is appropriate for a particular need?
- What are the problems?
- How much does it cost?
- Can I do it myself?
- What outside services are available?
- What are other organizations doing?

This book begins by answering the question "What is a teleconference?" and goes on to examine how teleconferencing developed, and to provide an overview of the field today. Chapter 2 presents a comprehensive description of the technology of teleconferencing. It covers every aspect of the underlying principles of audio conferencing, enhanced audio conferencing, one-way video/two-way audio conferencing, two-way video/two-way audio conferencing and computer conferencing. In Chapter 3, various issues and problems are discussed, including where teleconferencing fits in the organizational structure, what items a teleconferencing manager must consider and what problems he or she can expect. How, why and by whom teleconferencing is being used is the subject of Chapter 4, which reports the results of an exploratory survey conducted for this book.

Chapter 5 deals with the economics of teleconferencing. It gives actual and estimated budgets for many kinds of teleconferences and compares the cost of holding a face-to-face meeting. Chapters 6 and 7 examine the production of a teleconference, both in terms of the needs analysis that must be done beforehand, and the step-by-step production and implementation processes. Personal insights and hints on teleconferencing are provided by an experienced vendor and user, respectively. Detailed case studies of a variety of users—including Aetna Life and Casualty, Ford Motor Co. and United Technologies Corp.—are presented in Chapter 8, which discusses the spectrum of teleconferencing applications. Finally, the book includes a directory of audio and video conferencing suppliers, as well as a technical glossary.

The aim of this book is to provide the basic information needed to examine your own communications and meeting needs; to understand the teleconferencing modes available and how they compare with each other as to cost, effectiveness and practicality; and to take the first steps toward informed use of teleconferencing.

Ellen A. Lazer

1

Some Observations on Teleconferencing

by James W. Johnson

Teleconferencing is one of the hottest ideas to come along in many years. It has excitement, it's daring, it's a media frontier. It promises economy: money and manpower can be saved through proper use of teleconferencing. These are substantial and easily-grasped characteristics. Yet it is truly an enigma.

Teleconferencing has been defined as many things, some of which are contradictory: it is an entirely new concept/it is as old as television, or the telephone, or even the telegraph; it is simple/it is complex; it is inexpensive/it is very expensive; it must be interactive/it doesn't have to be interactive; and so on.

TOWARD A DEFINITION

What, really, is a teleconference? Since the earliest communications form of "tele" was telegraph, followed by telephone and television, it was to be expected that "teleconference" would follow. First used as a word in the 1960s by the U.S. Department for Defense Analysis to refer to computer conferencing, the roots of "teleconference" are *tele*, from the Greek meaning "from a distance," *com,* the Latin prefix meaning "together" and *ferre,* from the Latin meaning "to bring." Thus we have "bringing something together from a distance."

But to define a teleconference, I like to use the acronym "SPIES": A teleconference is a meeting that is Structured, Private, Interactive, Electronic and Scheduled.

- *Structure* implies an agenda, specified participants, and identified goals and expectations.

- *Privacy* is the natural result of custom networks and the desire to deliver a special message to special recipients. Whether a three-hour symposium to 15,000 doctors or an audio conference between two persons, privacy is always inherent in the structure of the event. The degree of privacy required may vary considerably: it may be "accidental," accepted just because it is an attribute of the medium, or it may be "essential," and necessary to the success of the conference.

- *Interactive* elements imply that the meeting is to some degree "live." A prerecorded program is not a teleconference, nor is a live program that lacks interaction. It is the live exchange of ideas that lifts the teleconference to exciting heights and provides the dynamics so essential to effective communication.

- *Electronic* describes the delivery system. Teleconferences involve telephones, computers, television cameras and the like—rather than media such as newspapers or interoffice memos.

- *Scheduled* events are planned, and it is this planning that differentiates a teleconference from a spontaneous telephone call.

Definition of a Teleconference
A structured, private, interactive, electronic and scheduled meeting between two or more persons in separate locations.

With this definition in mind, let's take a look at the development of teleconferencing. Later we will examine the reasons for its popularity, its modes and methods, and its future.

THE DEVELOPMENT OF TELECONFERENCING

Television's earliest strength and popularity stemmed from its extraordinary ability to bring the public an event "live," as it happened. This is the stuff of horse races and football games; people love to watch an event with an unknown outcome. Television has been able to link this quality not just to sports events but to "nonbroadcast" uses such as business meetings and surgery. This continues to be television's greatest value and teleconferencing's inherent promise. While such needs and events have long been with us, the tools of teleconferencing, described in detail in Chapter 2, are relatively new.

Kinds of Teleconferences

As the following chapter explains, there are other teleconferencing forms besides audio and full-motion video. A telephone conference call is an audio conference if it meets the aforementioned criteria of privacy, structure, schedule and live interaction. As such it has been available, and indeed used, for decades as a business tool. The introduction of the "meet me" dial-in conference bridge has greatly increased the medium's popularity and use. This technique requires only that the participants call a predetermined number, whereby they are "bridged" to each other. The conference bridge is sometimes used to provide the interactive two-way audio element of a video teleconference, where its asset is simplicity, though at the price of reduced control.

Slow-scan television transmits a still image slowly—in anywhere from 2 to 90 seconds and in a continuous spectrum of quality from poor monochrome to excellent color. The quality of the equipment and transmission lines used determines the speed and quality of the image. Equipment is generally available by purchase only, and is best used to transmit technical drawings and similar data of nonstandard size, which a TV camera zoom lens can easily accommodate. Transmission of standard printed matter is much better left to facsimile machines, which are already available in rapid transmission formats. Both facsimile and slow-scan TV signals can be sent over a standard telephone line, so by using one network for voice and another for data, an augmented, or enhanced, audio conference results. The electronic blackboard transmits images much as slow-scan does, but with a blackboard instead of a camera, and in real time.

Computer conferencing is really electronic message processing, and in my opinion is not properly teleconferencing at all (any more than letter-writing is "conferencing"), since it lacks the "live" element.

How Teleconferencing Grew

Teleconferencing owes its birth to what, for nearly three decades, was called closed circuit television. To understand that beginning, it is helpful to look at the evolution of other media.

Six centuries ago, books were available only to the elite. They were handwritten, expensive and scarce. Gutenberg's invention, movable type, signaled the beginning of a change that would give the print medium to the people. To all the people. Today, anyone can own a book. The first telephones were the prized possessions of a rare few. Yet, in only one century, we have reached the point where nearly every household in America has a phone, and everyone has access to one. The medium has been given to the people.

Television had a similar beginning. Initially, TV cameras were expensive, difficult to operate and few in number. But time brought the medium to the people—in this case, in only 35 years—and now anyone can own a camera and recorder for a few hundred dollars. And, with enough talent, anyone can produce a television program.

Today, we have teleconferencing, another unique medium. It, too, started out as a medium for the elite. It developed along two tracks, beginning in the late 1940s, at about the time television was introduced to the marketplace. The first track was medical, the second, box office.

Medical procedures were transmitted to meeting rooms, while sporting events were transmitted to theaters. In the field of medicine, CBS' field-sequential color system, while not compatible with the receivers of that time, was readily available and provided extraordinary color fidelity. Furthermore, it was ideal for large audience private TV, which requires three-color TV projection, an easily adaptable format. The Smith Kline & French pharmaceutical company (notable, among others) provided large-screen, color, interactive teleconferencing (then called closed circuit television, or CCTV) of medical procedures for the education of thousands of physicians at association meetings throughout the country. Meanwhile, box office sales for closed circuit prize fights were brisk, and by the 1960s World Cup soccer, bull fights from Spain, auto racing and a variety of theatrical events were transmitted for large-screen viewing. Through the years, these two tracks have grown.

Two people were central to the birth and growth of teleconferencing. In the late 1940s, William S. Paley (then president of CBS) had an assistant named Nathan L. Halpern, a recently discharged Navy ensign and law school graduate. Paley rejected the production of box office, closed circuit boxing matches, but Halpern saw its potential and left CBS to found Theater Network Television. Halpern's promotion and distribution of prize fights to theaters presaged the eventual multi-city closed circuit telecast and today's teleconferences. His company, later known as TNT Communications, transmitted the first international telecast by satellite, and the first major business (marketing) telecasts for such corporate giants as IBM, Ford, General Motors and Chrysler. Halpern's consistently high levels of production and logistic control set the standards for the industry that was to develop.

In the 1960s, Halpern was joined by Robert F. White, a young former actor who proved to have an unusual talent in the field of communications. First, White devised a multi-site closed circuit multi-media network, controlled from a single origination location. Called "Triggervision," it was one of the first truly innovative ideas to reach the closed circuit world. It was used by all the major auto manufacturers and today is controlled by a U.S. patent.

White's second contribution was the development of a single package of equipment designed exclusively for the video conference. He envisioned corporate use of this medium many years before it happened, and with his company, Management Television Systems, designed and produced over 40 packaged units consisting of a color TV projector with a built-in monochrome backup projector; a sound amplification system with giant speakers and high powered amplifiers; an 11x14 foot screen (the most common size); complete test and monitoring equipment; a remote control console; complete spare parts; and necessary tools. Designed for meeting rooms rather than for amphitheaters or movie houses, White's units proved to be successful teleconferencing tools, and were used by government and business long before the advent of the communications satellite.

White's design inspired the General Electric (GE) Co.'s Command Performance Network of the late 1960s. The Network never achieved commercial success, but it did result in the development of GE's popular large-screen TV projector, in common use today.

In recent years, cameras, recorders, transmitters and satellite receivers—all components of teleconferencing—have decreased in cost and increased in availability. Thus, a geometrically increasing population of communicators are finding new and increasingly effective ways to make video conferencing a part of our lives. Because earlier costs were high, the teleconference was reserved for major marketing events staged by the large corporations. Now, with reduced costs, routine "information" telecasts are the order of the day. The medium belongs to everyone.

THE MARKET FOR TELECONFERENCES

While the teleconferencing market is largely not understood and is poorly defined, it is clear to most observers that it is already large and promises to become enormous. Holiday Inn of America bet several million dollars on this with its HI-NET service, installed to provide in-room pay TV programming but expanded to include teleconferencing. Hilton Hotels and Marriott Corp. have followed suit on a more modest scale, and others are sure to follow.

Seeing millions of dollars of revenue within the near future, virtually all broadcasters, cable companies and common carriers have entered, or plan to enter, the teleconferencing business. The growth curve is steadily rising, and the base of teleconference users is constantly broadening, as shown in Figure 1.1. The period from 1950 to 1980 featured the introduction of many new uses for the teleconferencing medium—in medicine, sports, government, business, fund raising, to name a few. In the 1980s a few new applications will be found, but the existing applications will be greatly expanded. Note the curve's slope in the 1990-2010 period. Those who think teleconferencing's growth is spectacular will be astounded by the future.

Major Users and Applications

Typical users of video conferencing include:

- Major manufacturers (e.g., Xerox Corp., 3M Co., IBM, Firestone Tire & Rubber Co., Ford Motor Co., J.C. Penney) motivating or informing their customers or their employees;

- Government agencies (e.g., Environmental Protection Agency, Occupational Safety & Health Administration, Department of Commerce) promulgating essential, changing data;

- Health care businesses (e.g., Roche Laboratories, Meade Johnson & Co., Ortho Corp., Syntex Co.) bringing new topics to physicians;

- Fund raising organizations (United Jewish Appeal, Republican National Committee) linking fund-raising dinners by television;

- Trade and educational institutions (e.g., the American Bar Association) offering symposia for a fee;

- Religious organizations (e.g., the Roman Catholic Church) bringing worship-oriented events to private audiences throughout the country.

As is described in Chapter 8, teleconferences are also used for press conferences, e.g., the introduction of feature films or other products to the press, and fashion publicity, bringing Paris fashions to Chicago buyers. The uses are virtually unlimited.

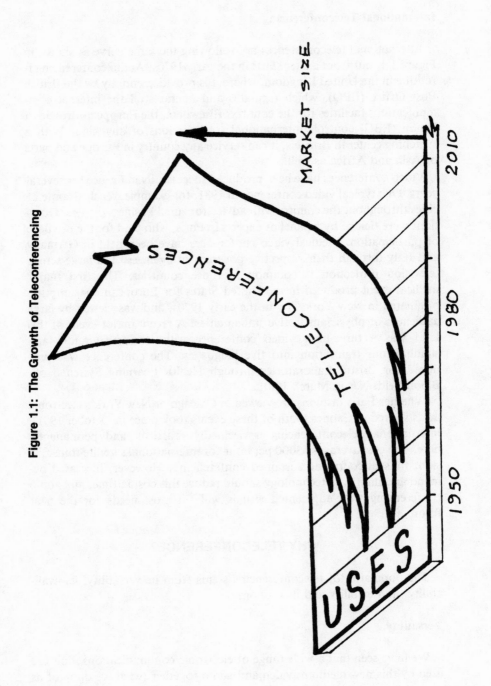

Figure 1.1: The Growth of Teleconferencing

International Teleconferencing

International teleconferences are following the same curve as shown in Figure 1.1, but from a later start: in the early 1970s. Audio conferencing is routine in the United Kingdom, where it is provided entirely by the British Post Office (BPO), which regulates and operates all the internal communications facilities for the country. Elsewhere, the European Broadcast Union (EBU) controls international transmissions of any kind. With a switching center in Brussels, it can service any country in Europe and parts of Asia and Africa as well.

Video conferences have been produced successfully in France for several years. One typical video conference in 1981, for example, reached some 11 cities throughout the country with advice for small business owners. Questions were fielded by a panel of experts in Paris, who paid for the privilege of participation. Medical video conferences have been held in Germany and Italy through their respective government agencies, and large-screen television projection is common in those countries. The first major medical event produced in the United States for European consumption originated in New York City in the early 1970s, and was viewed by large audiences of physicians in five Italian cities. A recent major event of this kind was a three-hour video conference with interactive audio and simultaneous translation into five languages. The conference was produced for Bristol Laboratories through Health Learning Systems, of Bloomfield, NJ, in March 1981.

Whether Paris fashions are viewed in Chicago or New York's electronic art is viewed in Cannes (both of these events took place in October 1981), international teleconferencing is decidedly practical, and permanently here. The cost averages $3000 per hour for international satellite transmission, thus discouraging sustained conferencing. However, increased demand and improved technology should reduce this cost in time, and audio conferencing with augmented visuals will fill most needs for the near future.

WHY TELECONFERENCE?

The popularity of teleconferencing stems from its versatility, its availability, its flexibility and its economy.

Versatility

We have seen that a wide range of electronic communications tools are used by this new medium: video and audio together (what we think of as television); audio only (the "conference call"); audio with still-frame or slow-scan video; audio with facsimile; audio with local "in situ" visuals (slides, film, video tape, drawings, etc. can be shown locally, illustrating a telephone conference); and audio with certain other devices (such as the electronic blackboard).

Availability

This is illustrated by the ease with which nearly anyone can teleconference today. Telephones are everywhere. TV production services are offered by hundreds of suppliers and TV equipment can be purchased or rented from local dealers. There are thousands of permanent earth stations owned by cable stations, hotels (Holiday Inn's HI-NET features over 300 locations), government and private companies. Mobile TVROs (receive-only satellite earth stations) are steadily decreasing in cost and can be rented from dozens of suppliers. Uplink service is available from the major carriers (Western Union, RCA, AT&T, Intelsat), the Public Broadcasting System (PBS), privately owned mobile transmitters and from the Public Service Satellite Consortium (PSSC). (See the Directory at the back of this book for a listing of these and other organizations.)

There is practically no place on earth from which one cannot originate a teleconference, or receive its program. Costs are variable, however, and prices rise as accessibility declines, as Chapter 5 shows. For example, transmission cost from New York City's WNEW-TV studio is a minimum $620 plus $99.10 per hour to reach a local uplink site, while a recent telecast from a Westchester hotel (located in a valley 20 miles north of New York City) incurred a whopping $13,000 in special microwave installation to reach the same site. Similarly, on the receive side, a mobile TVRO is available for around $1000, while microwave service alone to a downtown Dallas meeting site, deep in the Stone Canyon, was recently priced at $8000.

The availability factor also refers to the availability of trained personnel. Personnel to operate equipment and produce programs are essential, but are currently in short supply and mostly inexperienced. This is being ameliorated, however, with many colleges offering communications training. By 1995, virtually all executives will have grown up after the introduction of TV, and will have had at least some formal communications education. Their familiarity with the media, coupled with an intimate relationship with computers and video games, will result in routine, informal use of the teleconference. The electronic communications medium will, indeed, be theirs.

Flexibility

As a communications medium, the teleconference is also tremendously flexible. It effectively fills an unusually wide variety of needs in marketing, education, information dissemination, meetings, entertainment, government and fund raising. Many for-profit educational events, such as TeleSeminars,* are popular today, and there will be additional applications in the near future. The recently introduced PressNet,† a low-priced electronic press conference, is typical of these, as is Speaker By Satellite,† which brings guest speakers to distant meetings.

Not only does teleconferencing fill many needs, it adapts to almost any business. As the previous examples indicated, automotive and computer manufacturers, retailers, government agencies, business associations, pharmaceutical manufacturers, nonprofit groups, religious organizations and many others have all benefited from the use of the teleconference, using combinations of all teleconferencing modes, from audio only through full motion video.

REASONS FOR THE GROWTH OF TELECONFERENCING

The extraordinary growth of the teleconference can be attributed to technological, economic and psychosocial factors. For an understanding of the medium, it is helpful to examine each of these.

Technological Factors

These factors are best understood by examining the various parts of teleconferencing as they are affected by ever newer electronic developments. Cameras and recorders are better, cheaper and easier to operate with each growth stage. Transmission, control and display technologies are burgeoning. This is popularizing the design and installation of the permanent electronic conference center or room, eventually to become a key facility of every major business, association and government office.

While at the time of this writing the cost of teleconferencing is compared with meeting expenses in order to be justified (and the monies borrowed from travel budgets), soon these comparisons will decrease in importance and the medium will have a rationale (and a budget) of its own as an essential business tool. The ready availability of inexpensive transmission methods is producing an important shift from marketing to informational events and to ever growing numbers of permanent conference rooms.

---

*TM—VideoNet, Inc.

†TM—TeleConcepts in Communications, Inc.

AT&T's Picturephone® Meeting Service (PMS) has permanent conference rooms available for use on an hourly basis, on short notice.

The Darome Connection, one of the first telephone conferencing independents, has been joined in the marketplace by Connex International and a number of other telephone conferencers. Slow-scan manufacturers such as Colorado Video are providing better and less costly equipment for an increasing number of industrial users. In short, the energy once reserved for the development of TV itself is now being focused on teleconferencing, which is being seen as a truly exciting growth opportunity—and perhaps the last great communications frontier.

Economic Factors

The economic impact of teleconferencing has been the subject of much popular and somewhat superficial discussion. People see only the surface costs and benefits—they should also examine manpower costs, costs of not holding a meeting, lost opportunities, and so on. Does your company presently do a touring "dog and pony" show? Add the cost of travel, lodging, personnel, time and the multiplicity of meeting components and compare them to the cost of a teleconference. Of course, just looking at these savings does not give a true picture. (Chapter 5 provides true comparisons for assessing teleconferencing, touching on risks and benefits, costs and savings.) However, whether real or merely perceived, the economic factors have a very strong influence on the increasing use of the teleconference.

Psychosocial Factors

The psychosocial elements of teleconference's growth are particularly interesting. The easy familiarity people today have with the telephone will be transferred to nearly everyone involved with teleconferencing. The impact on business of simple, reliable communications, available on many levels, will be immense. Teleconferencing provides a practical application of the electronic communications skills being learned by today's—and especially tomorrow's—businesspersons. Shifting from travel to teleconferencing will be easy, since travel is really a communications medium. We travel in order to communicate. As new avenues of communications open, travel communications will decrease. By the early 2000s the "cottage office," where people communicate (i.e., work) from their homes, will be commonplace. Actual physical travel will then be less necessary. Where now our natural medium is face-to-face, and the teleconference is the specialized alternative, tomorrow those positions will be reversed, and in-person meetings will be reserved for special events.

CONCLUSION

Teleconferencing will be a daily part of our business and eventually of our personal lives. In due course, it will lose its singularity. Teleconferencing "companies" will dwindle. As we don't "produce" a letter or a telephone call today, we won't "produce" teleconferences tomorrow. Only specialized "marketing" events will be produced in this sense. Just as when we talk to someone today, we don't consciously think of ourselves as "communicating," so will tomorrow bring a greatly lessened consciousness of the act of "teleconferencing." It will become a routine tool within a score of years for all who make effective communications a part of their lives.

2

The Technology of Teleconferencing

by Martin C.J. Elton and David Boomstein

INTRODUCTION

Teleconferencing is generally defined as the use of telecommunications systems to enable three or more people, at two or more locations, to confer with one another. Some argue that the definition should be broadened to include two-person interactions. Still others would replace the term "confer" with another, which covers one-way transmission of information as well as two-way communication.

The Choices Available

Today, three very different kinds of activities are described as teleconferencing. First, there is the electronic analog of the business meeting. Here the technological design values tend to emphasize the importance of interaction. Second, there are what may best be termed performances: business seminars, public relations events and so on. Here the production values are generally reminiscent of broadcasting. Finally, there is the process of communication extending over periods of weeks or months, and intermingled with a variety of other ongoing processes. Computer conferencing, as this process is called, employs computers to store messages transmitted from a keyboard until the intended recipients log in to collect them.

For the purpose of this book, we will describe teleconferencing in terms of the media which are employed. As shown in Figure 2.1, five variations can be identified:

Figure 2.1 The Five Main Types of Teleconferencing

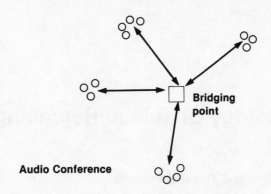

Audio Conference

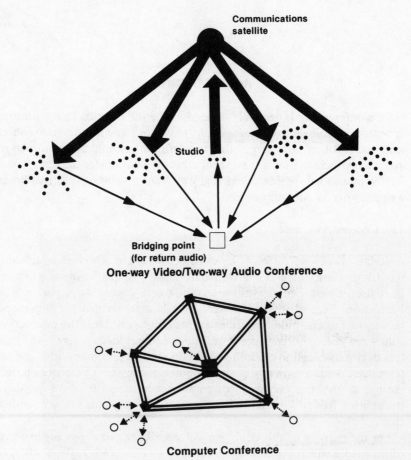

Computer Conference

Figure 2.1 The Five Main Types of Teleconferencing (cont.)

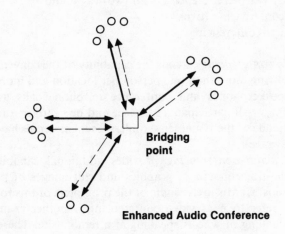

Bridging point

Enhanced Audio Conference

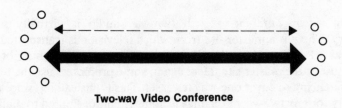

Two-way Video Conference

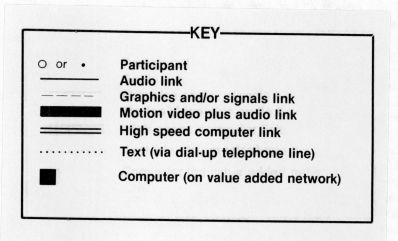

KEY	
O or •	**Participant**
———	**Audio link**
– – – –	**Graphics and/or signals link**
▬▬▬	**Motion video plus audio link**
═══	**High speed computer link**
··········	**Text (via dial-up telephone line)**
■	**Computer (on value added network)**

- Basic audio conferencing
- Enhanced audio conferencing
- One-way video conferencing with two-way audio
- Video conferencing (two-way)
- Computer conferencing

Basic audio conferencing extends the capability of the conventional telephone by allowing more than one person per location and more than two locations. Speakerphones and conference telephone calls are its best known examples. All participants can hear and be heard by one another, but no use is made of the visual sense (unless visual materials have been circulated in advance).

Enhanced audio conferencing provides additional capabilities. One example is the transmission of graphics and still images of participants among locations. Confusingly, some of the promoters of this form of teleconferencing refer to it as video conferencing. Another example is the automatic signalling of who is speaking at a remote site. There are other possibilities, too, such as remote drawing devices (e.g., electronic blackboard).

One-way video conferencing with two-way audio is virtually self-explanatory. There is a master site from which television is transmitted to all other sites. Participants at these other sites can speak with, but not be seen by, those at the master site. Here again, some promoters use the term video conferencing to cover one-way telecasts. This is misleading, since the opportunity for any two-way exchange is nonexistent, or limited to dialing a toll-free telephone number if one has a comment or question.

Two-way video conferencing allows all participants to see and be seen by, as well as hear and be heard by, one another. At present, business video conferencing is generally limited to two sites.

Computer conferencing blends into a family of emerging telecommunications services known as *electronic mail*. The emphasis, however, is on group communication, rather than one-to-one communication. Use is made of *value-added networks* (VANs), electronic superhighways linking together computers, which provide for highly economical long-distance transmission. Communication takes the form of text entered at a keyboard by the originator, and displayed on a screen or printed on paper for the recipient. The input-output equipment is a computer terminal or microcomputer, usually connected via the regular telephone network to the nearest point on the appropriate computer network. Communication is rarely in real-time. More commonly it is asynchronous—i.e., messages are held in the computer until intended recipients check in.

Computer conferencing has been described here in order to give a complete picture. Since this book deals with the real-time uses of audio and video teleconferencing, we will not go further into computer conferencing technology. It is worth noting, however, that there is a sense in which all the teleconferencing media are synergistic. *Combinations* of several or all of them may enable new organizational configurations (either centralization or decentralization) or new relationships between the locations of one's home and one's employer—for example, the electronic cottage. In these cases, the availability of one form of teleconferencing may enhance the value of others. The National Aeronautics and Space Administration (NASA) is a good example of an organization employing a judicious use of a variety of teleconferencing media. Audio, enhanced audio and two-way video teleconferencing have been used for a variety of activities, particularly for coordinating geographically dispersed contractors.

Table 2.1 summarizes some key points relating to the four teleconferencing media with which this chapter is concerned. Notice that there is no relationship between bandwidth (the capacity for carrying information) and flexibility. If one needs equal participation at three or more sites, or if one needs to involve overseas locations, basic or enhanced audio conferencing is the best game in town.

Structure of This Chapter

This chapter describes the technological options available for audio and video conferencing, and provides an explanation for the layperson of how the different technologies work. Since this is a fast-moving field, we will emphasize the underlying principles, rather than describe in detail the features and prices of individual competing products.

The following section provides an overview of some basic principles of teleconferencing technology: What complications arise as one moves from telephone conversation to audio conferencing, and from enhanced audio conferencing to either form of video conferencing? What are the major subsystems of a teleconferencing system? What is the difference between analog and digital transmission, and when does it matter?

Next we turn to audio conferencing, considering the end-equipment (equipment located at the site), the conference rooms, the lines over which the electronic signals are transmitted and the way these lines are connected together. This is followed by a description of the various enhancements that are potential adjuncts to audio conferencing: freeze-frame (or slow-scan) television, graphics tablets and the electronic blackboard, remotely controlled projection equipment and means of speaker-identification.

The last section covers the two forms of video conferencing. It deals with end-equipment, conference rooms and transmission options.

Table 2.1   Comparative Features of Teleconferencing Systems

Characteristic	Basic Audio Conferencing	Enhanced Audio Conferencing	One-way Video Conferencing	Two-way Video Conferencing
Maximum number of locations	Usually from 16 to 50, depending on bridge	Usually from 16 to 50, depending on bridge	Usually as for basic audio conferencing	Usually only 2
Typical number of locations	From 2 to 6	From 2 to 6	Highly variable, depending on purpose	2
Maximum number of persons per location	A few hundred, if lines allowed at microphone	A few hundred, if lines allowed at microphone	A few hundred	From 6 to 12
Typical number of persons per location	Up to about 5	Up to about 5	Highly variable; usually from 20 to 200	About 5
Fully two-way transmission	Yes	Usually	No	Yes
Locations outside North America	Easy	Easy, if terminal equipment available	Very unusual and expensive	Very unusual and expensive
Number of technical/production personnel	Usually at most 1 (at bridge)	Usually 1 at bridge; maybe 1 per site	Usually at least 6 at center and 1 per remote site	Usually 1 on call at either end
Relative average cost per occasion	Low	Fairly low	High to very high	High to very high
Relative average cost per participant	Low	Fairly low	Fairly low to moderate	High to very high
Constraints on location of sites	Often only access to telephone network	Where terminal equipment is located	Large enough rooms; access to satellite dishes; availability of rooms	Usually one of a fixed set of installed rooms

GENERAL CONSIDERATIONS

From Telephone to Audio Conferencing

The telephone handset, for all its seeming simplicity, is a masterpiece of design. Because the microphone is so close to one's mouth, the direct energy it receives from one's voice vastly exceeds the energy it receives from other sources—e.g., the voice reflected from hard objects (such as the walls) and ambient noise. Because the speaker is so close to one's ear, the volume can be low enough to prevent sound from leaking through the air from the earpiece to the mouthpiece.

But what happens when there are several participants in a conversation at one location? They cannot use parallel extensions (two or more telephones on the same number) because this causes an unacceptable weakening of the incoming and outgoing audio signals. Besides, holding a telephone handset for an hour is decidedly uncomfortable. Therefore, microphones and loudspeakers are shared among the participants. This gives rise to two problems. First, microphones are much farther from peoples' mouths, so the ratio of direct voice energy to reflected energy and noise is much less favorable. As a result, voices may sound hollow and indistinct. Second, incoming sound must be at a higher volume in order to be heard by ears that are no longer next to speakers. As a result, the sound can leak back into the microphones, causing the "howlaround" experienced with poorly set up public address systems. Carefully designed room acoustics and sophisticated electronic circuitry are needed to overcome these two problems.

Another type of difficulty occurs when more than two locations must be connected. Connecting the corresponding pairs of wires, as amateur electricians might do at junction boxes in their homes, is not sufficient. Some telephone channels may be carrying a strong signal and others a weak signal; the variability is quite substantial. Such differences result in unacceptable audio quality. Additionally, a certain level of noise will be transmitted from any location, regardless of whether someone there is speaking. All such noise is additive and can start to overwhelm the intended signal, unless special electronic measures are taken. To overcome these and other transmission problems, channels are interconnected at specially designed *bridges* (discussed later in this chapter).

From a technological point of view, the need to accommodate more than two locations and more than one person per location creates problems requiring considerable technical skill, if not virtuosity, to solve. Fortunately, these problems have been receiving due attention in recent years, particularly (as is appropriate) within the Bell System. Some outstanding equipment, as described later in this chapter, is now coming onto the market.

From Audio Conferencing to Video Conferencing

From the perspective of teleconferencing there are two crucial differences between visual and aural information. First, just as the rate at which water can flow through a pipe is limited by the diameter of the pipe, so too is the rate at which information can be transmitted via a telecommunications channel limited by the *bandwidth* of the channel. Unfortunately, the bandwidth required just for a black-and-white television channel (yielding the same resolution as broadcast television) is approximately one thousand times the bandwidth required for a telephone channel. Because there are economies of scale, the ratio for transmission cost is much better than 1000 to 1, but a ratio of even 50 or 100 to 1 makes the transmission of television much more costly than that of the telephone.

As we will discuss later, there are ingenious means by which the bandwidth necessary for video conferencing can be reduced substantially. Nevertheless, the terminal equipment needed (codecs, discussed in the "Video Conferencing" section) represents a high fixed cost. It only becomes economical if use is frequent enough.

The second difference between visual and aural information is so elementary that it is often forgotten: sound is additive but visual images are not. Within broad limits, we can hear sounds, coming simultaneously from many different sources and directions and still understand what is being said by any one source. In a disciplined audio conference, in which only one person is speaking at a time, we should experience no difficulty hearing, provided the technology is working properly. Sight, however, is directional, and if visual images are superimposed one upon another (on a screen), it is usually impossible to see what is happening.

As a result, if we need a constant visual image from each location in a video conference, then we need a separate incoming channel and a corresponding screen for each location (or, possibly, a split screen). In total, the number of one-way channels rises almost as the square of the number of locations (the actual formula is n^2-n, where n is the number of locations). The corresponding costs are generally prohibitive.

A more economical alternative would be to limit the number of locations from which visual images are transmitted. For example, if the person speaking was at site D, then the people at sites A, B and C could see site D, but not each other, while the speaker at site D could see site A, B or C, depending on who spoke last. However, this is still expensive: two-way channels are needed between each site and a central switching location, and automated switching cannot be provided cheaply. Furthermore, it is usually impossible to receive a visual cue that someone at another site wishes to interrupt, since automatically switched systems are reactive, not anticipatory.

The third possibility is to substitute a person, or persons, for the automated switching system. This imposes labor costs and sacrifices the confidentiality which may be important in a business teleconference.

At present, there is no one commercial offering that provides an economical two-way video conferencing system for business purposes, among three or more sites. For multi-site video conferences, the video can only be transmitted from one site. For two-way video conferencing, only two sites can be connected. This problem has been solved in various *ad hoc* ways, primarily in the fields of telemedicine and educational television.

The Interactive Hospital Network in Hanover, NH, has demonstrated a successful five-way audio/video conferencing network. Also, economical solutions have been demonstrated for video conferencing via cable television channels, e.g., the senior citizen's system in Reading, PA and the school district system in Irvine, CA. However, in these examples, neither bandwidth nor labor imposes a significant cost—many cable TV systems have unused channels which they make available, and the labor involved is often volunteer. Other *ad hoc* systems include the Defense Advanced Research Projects Agency, which has had a prototype system constructed to interlink four sites. (Cost-effectiveness, however, must be construed in a different way in the field of defense operations.) Also, AT&T did offer three-point video conferencing among certain locations in the Picturephone® Meeting Service (PMS) trial which ended in 1981. However, this is not a feature of the present offering.

Telecommunications Subsystems

For descriptive purposes, a telecommunications system is usually divided into two or three component subsystems whose purposes are *transduction, transmission* and, in telephony but not in broadcasting, *switching*.

Transduction

The transduction subsystems change information from the form in which it is recognizable to our senses into electronic form, and back again. Thus, microphones convert air pressure waves (i.e., sound) into waves of electrical current; loudspeakers reverse the process. Television cameras convert light waves into electrical current; television monitors reverse the process. The transducers are essentially the terminal equipment used for teleconferencing.

Transmission

Once information is in electronic form, the transmission subsystem comes into play. Information can be transmitted in many different ways. It can be carried through the air by electromagnetic waves, a method employed by microwave relays and communications satellites. It can be carried along wires, either the twisted pair, which is the usual way for telephone lines to enter users' homes and offices, or coaxial cable, which is used for cable television systems. In the near future we are likely to see increasing use of optical fibers (hairlike strands of glass or plastic); in these, information is transformed into patterns of light waves, which are generated by lasers and then transmitted through the fibers.

Often the communications channel combines different physical carriers. Thus, a long distance telephone call via a communications satellite still makes use of twisted wire pairs (at either end of the connection).

This is a convenient point at which to make a few general remarks about transmission, which are discussed more fully later in this chapter. First, the various transmission carriers differ considerably in the bandwidths provided. A twisted wire pair (narrow bandwidth) is inadequate for carrying television, whereas an optical fiber (wide bandwidth) can carry hundreds of television channels. Second, carriers differ in the relationship between cost and distance. The costs of communication satellites do not depend upon distance, provided only one hop is needed, whereas the costs of transmission by cable are approximately proportional to distance.

Third, when considering the economics of long-distance transmission, it is often necessary to consider two components: the short-haul costs at either end (i.e., the "last mile costs") and the trunk transmission costs in between. The plant used for trunk transmission is usually much more intensively utilized, since it is shared among many users; hence the corresponding costs per mile are generally much lower. For the public-switched telephone network there is another reason for distinguishing between the short-haul and long-haul components. As shown in Figure 2.2, the network is hierarchical. When comparing the transmission quality between the higher level switching centers in the network with the transmission quality in the local loop, one finds that the former is much better. Last, signals of different kinds may sometimes be combined, or *multiplexed,* for transmission along a single channel with separation at the receiving sites.

Switching

A switching subsystem is necessary when connections are to be made among certain specified sites and no others. In the telephone network, this

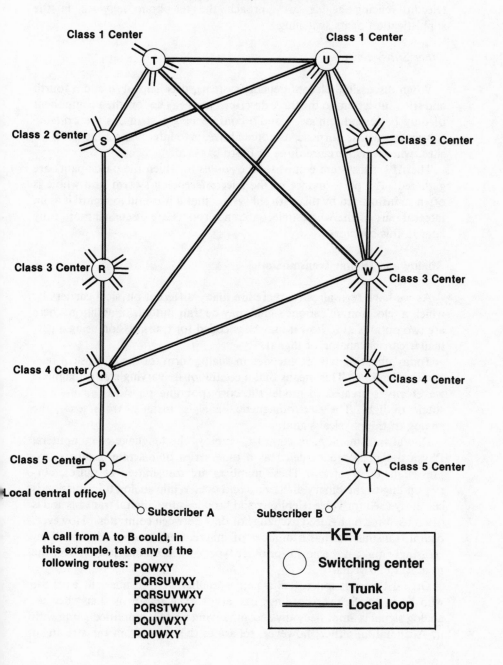

Figure 2.2 The Hierarchy of Switching Centers in the Public-switched Telephone Network

Class 1 Center

Class 1 Center

Class 2 Center

Class 2 Center

Class 3 Center

Class 3 Center

Class 4 Center

Class 4 Center

Class 5 Center

Class 5 Center

(Local central office)

○ Subscriber A

Subscriber B ○

A call from A to B could, in this example, take any of the following routes:

PQWXY
PQRSUWXY
PQRSUVWXY
PQRSTWXY
PQUVWXY
PQUWXY

—KEY—

◯ Switching center

—— Trunk

══ Local loop

subsystem incorporates the dialing mechanism, equipment in switching centers and the ringing mechanism. No other transmission system used for teleconferencing begins to approach the telephone network in the sophistication of its switching.

Other Subsystems

When discussing teleconferencing systems, it is helpful to add a fourth and fifth subsystem to the three described above. The fourth, a component of only basic and enhanced audio conferencing systems, is the *bridging* subsystem, whose purpose, as noted above, is to interconnect the different sites, whenever there are three or more of them.

The fifth subsystem comprises the *rooms* in which the participants are gathered. The performance of the teleconferencing system as a whole is often so influenced by this fifth subsystem that it is useful to regard it as an integral part of the whole (teleconference rooms are discussed more fully later in this chapter).

Analog and Digital Transmission

As we have seen already, there are many different physical carriers by which a telecommunications signal may be transmitted. In addition, there are two options as to how it may be encoded for transmission using a particular carrier: analog or digital.

Today, most signals are encoded in analog form for the purpose of telecommunications. This means that a continuously varying electromagnetic waveform is created to model the corresponding physical signal—e.g., sound or light. The electromagnetic signal is, in an obvious sense, the analog of the physical signal.

Digital transmission, in contrast, employs the language of computers. Physical signals are modeled as a time series of discrete numbers, expressed in binary form. These numbers are transmitted as pulses, corresponding to the binary digits zero and one. While analog transmission is currently used for voice telephony and broadcasting, digital transmission is often used for high-speed exchange of data between computers. However, digital transmission has a number of inherent advantages which will increasingly make it the norm for other types of telecommunication over the next 10 to 20 years.

Digital transmission offers a higher quality of transmission, as those who listen to digital recordings can attest. The reason is that when an analog signal is amplified, any accompanying noise is amplified along with it. A digital amplifier, however, separates the noise from the stream of zeroes and ones, and transmits only the zeroes and ones. Consequently, noise is not amplified in digital transmission.

Another advantage is that computers, because they share the same language, can economically process digital transmissions. This potential can be used for encryption, which provides a higher degree of security than conventional scrambler devices. More important, as far as most video conferencing users are concerned, microprocessors can be used to "compress" television signals by removing redundancy and thus reducing the amount of bandwidth necessary. (Redundancy is the use of more information than is necessary to encode a signal.) The two primary forms of redundancy are 1) information which is the same over a wide area within a single television frame (e.g., an area of wall) and 2) information which remains the same between successive frames (e.g., stationary objects as viewed by a stationary camera). It is much more efficient, for example, to encode the fact that a specific large area is the same as in the previous frame, than to encode over again what each of the large number of cells comprising that area looks like. Given the high costs of television transmission, *bandwidth compression* (discussed in more detail under "Video Conferencing") is a very attractive feature of the digital mode.

There must be a match between the transmission and bridging subsystems, on one hand, and the digital or analog form of the signal, on the other. In other words, an analog signal must be converted to digital form before it can be sent along a physical channel designed for digital transmission. Sometimes the conversion must be made in the other direction, as when a computer terminal is connected to the regular telephone network.

The needs of computer users have spurred the increasing availability of digital channels of various information carrying capacities—capacities in excess of that of the analog telephone network. These channels may be used for high resolution graphics, which would have an impact upon enhanced audio conferencing, as well as for two-way video conferencing.

AUDIO CONFERENCING

This section considers the various technological components that comprise basic audio conferencing, from the audio terminal equipment needed to various transmission and bridging systems.

Audio Terminal Equipment

Except when there is only one person at a location, a conference telephone set must be used. Any conference telephone set includes one or more microphones and one or more loudspeakers. It may provide outlets for additional microphones and for additional speakers, e.g., for connection to a public address system.

There are more than 10 U.S. suppliers of conference telephone sets, in addition to the Bell System. Among them, they have about 40 different products on the market, ranging in price from $30 to $2000. The following are the more important respects in which the products differ:

1) Whether they are portable or for permanent installation in regular offices or special conference rooms.

2) The maximum size of group for which they are intended, which generally falls in the range of from 5 to 20. (The number depends on the acoustic characteristics of the rooms in which they are used, discussed in the next section.)

3) Whether they are complete units or adjuncts. The former include their own dials and ringers. When transmission is via the public telephone network, the latter require the separate use of regular telephone instruments to establish the calls.

4) Whether and how they employ *voice-switched loss* in their circuitry, the subject to which we turn next.

Voice-switching

The purpose of voice-switched loss is to reduce or eliminate the signals transmitted from all microphones other than the one into which a participant is speaking. When someone speaks loudly enough to override an incoming signal, his or her microphone is automatically switched from a receive mode into a transmit mode. This reduces the transmission of ambient noise and avoids sound from a loudspeaker reentering the transmission circuit. The latter is a cause of echoes and audio feedback ("howl-around").

In solving some problems, early voice-switching systems often introduced others. Users would sometimes complain about difficulty in capturing the circuit and about clipping, i.e., the loss of the first syllable of a remark. (If a switch does not operate fast enough, there is clipping. However, if it is too sensitive, it may cause instability. The design problem is far from trivial.) Fortunately, the state of the art has improved considerably in recent years and there is better understanding of how room acoustics and transmission levels help to prevent these problems.

There are two alternatives to voice-switching. One is the use of a push-to-talk control, which activates the corresponding microphone. Equipment adopting this approach is inconvenient to use in highly interactive teleconferences, but is widely used in teaching situations.

The other possibility is to have *open audio*. In this case, all microphones are live all the time. This is feasible if each user has a headset, as with a system successfully marketed by the Swedish telecommunications administration for some years.

Open audio may also be feasible when four-wire transmission circuits are used instead of the dial-up telephone network or some other two-wire transmission system. (These distinctions are explained below.)

Microphones

The proper selection of external microphones can make an important difference in some situations, especially when there are large numbers of users at a location or when it is important to give users as much freedom of movement as possible. Some systems have only a built-in microphone and offer no choice—e.g., the Bell System's familiar 4A Speakerphone. Some systems have a built-in microphone and can also accommodate two external microphones—e.g., Bell's 50A Conference Telephone Set and Precision Components' 50B Conference Telephone Set. With these systems, it is better to use only the external microphones if possible. There are also systems which provide for multiple external microphones.

The most frequently used external microphones are probably omnidirectional microphones placed on conference tables and lavaliere microphones hung around the neck. Dedicated (permanent) conference rooms often employ table-mounted microphones. In addition, two other possibilities should be mentioned, both of which employ directional microphones.

The Bell System introduced the Quorum® microphone in spring 1982. This microphone is designed to be highly sensitive within the vertical range above a conference table up to participants' mouths, and insensitive outside that range. It provides superior pick-up quality up to a distance of about 12 feet in a quiet conference room.

In a conference setting, shotgun microphones, which are highly directional, have proved to be worth the extra cost. They avoid the need for questioners to line up at a microphone and eliminate the pauses while questioners make their way to microphones.

The practitioner who is not used to audio conferencing should note two tips. First, if you want to try out a conference telephone set, the most important consideration is what it sounds like from *another* location. Have someone you know speak to you through it at a number of different distances from the microphone. Second, the correct connection of a conference telephone set to a public address (PA) system requires expertise. If the expertise is unavailable, better results will probably be obtained by placing a PA microphone next to the conference set's loudspeaker, rather than by attempting an electrical connection.

Conference Rooms

Users may take part in audio conferences from their own offices, from specially equipped conference rooms or from *ad hoc* conference rooms (e.g., in hotels). There are many other possible locations too—homes, pay telephones, classrooms, courtrooms and so on—but they do not raise issues which need detain us here. Whether an audio conference room is required depends primarily upon the number of participants involved and the need to use equipment which is expensive or not portable. Most probably this would be equipment for enhanced audio conferencing, such as that described in later sections of this chapter.

The acoustic properties of conference rooms are important. As these properties deteriorate, active participants must be closer to microphones if they are to be heard distinctly at the far sites. The maximum acceptable distance between a participant and a microphone is determined by the reverberation of a room and the ambient noise in it. Unless directional microphones are used, this is generally between 1 and 2½ feet.*

Long-term reverberation, or echo, will be less if the room is smaller and opposite walls are not parallel. Short-term reverberation causes a hollow quality in the transmitted sound which, when severe, is known as the "rain barrel" effect. It will be less if there are large irregularities, such as recesses and pillars, to scatter the sound, and if there is acoustic treatment to absorb sound. Finally, ambient noise will be less if a quiet room (e.g., an interior one) is chosen in the first place; if internal sources of sound, such as noisy air-conditioning equipment, are replaced or removed; and if acoustic treatment, such as double-glazing of windows, is used to prevent penetration of noise from outside.

Other aspects of room selection and design for audio conferencing are usually straightforward. Layout only becomes an issue when cameras are used for sending *freeze frame* images of participants (discussed later). In this case, seating should ideally be such that the camera can capture at least a three-quarter face view of any participant.

Ad hoc audio conference rooms have been set up with success at professional meetings in order to "bring in" remote panelists. In such cases,

---

*The Acoustics of Teleconference Rooms, AT&T, Technical Reference, PUB 42901 and 42903, explains how reverberation and noise may be measured and used to determine this distance.

there are two priorities: to select the room well in advance (to avoid the horror of a large space, with flat bare walls and hard floors, and a high level of external noise) and to ensure that the necessary telephone lines will be available. This often means installing lines which bypass the private exchange. (Private exchanges are automatic exchanges, today's successors of the manual switchboards. They often reduce transmission quality.)

Transmission

There are three main types of analog transmission systems used for audio conferencing:

- Public-switched telephone network
- Private lines dedicated to teleconferencing use
- Private networks installed primarily for regular voice telephony

The regular "dial-up" telephone network, known as the message telephone system (MTS), is most frequently used. Dedicated two-wire and four-wire telephone circuits are sometimes leased from the telephone company so as to be used exclusively for audio conferencing. Some organizations use their private telephone networks for audio conferencing, as well as other purposes. (Often these networks employ additional equipment to enable them to spread the available bandwidth over more telephone conversations than would be possible if the regular telephone bandwidth—3 kilohertz—were devoted exclusively to each conversation. This usually involves a trade-off between cost and quality.) In terms of performance and cost criteria, the most important respects in which the transmission systems differ are transmission quality, accessibility, the type of bridging they permit and cost.

The key aspect of transmission quality is its *signal-to-noise* ratio, essentially its clarity. Another relevant aspect is the *loss* in any line. This is measured in *decibels,* units of a logarithmic scale by which the volume of sound is measured. Variation in line loss during the course of an audio conference causes problems because, ideally, audio volume (the line level) should be the same for the signal coming in from any site. Dedicated circuits give the best transmission quality and, generally, multipurpose private networks give the worst (unless a premium service has been purchased). The MTS comes in between. Its main problem is variation in line level, but this can be largely overcome by sophisticated conference bridges. With a good bridge (and good terminal equipment), excellent quality can be achieved with the MTS, even in conferences involving several overseas locations. The occasional bad connection can be recognized at once and the party concerned can redial (or be redialed) before the conference begins. (Areas with inferior telephone service rarely feature in teleconferences.) The delay which is present when circuits on communications satellites are used is rarely a significant problem.

By accessibility, we mean the ability to involve participants regardless of their location. Evidently, the MTS, with its almost universal penetration, is close to ideal. Dedicated circuits are worst, since only the sites which are on the dedicated system can be connected. However, with a suitable bridge, one can use dedicated lines for those locations most frequently involved and the MTS for others. Multipurpose private networks vary enormously in their penetration, as each is custom made.

As will be discussed in the following section, there are various ways in which bridging may be provided. With the MTS, all options are open. Otherwise, one must install and operate one's own bridge.

The cost variations are obvious in general terms, though actual costs depend very much on geographical factors and utilization. Multipurpose private networks are the cheapest (assuming that they are sensibly designed). If usage is low, the MTS is next lowest in cost. However, there are generally cross-over points above which dedicated circuits are most economical.

The foregoing differences among transmission systems are summarized in Table 2.2. This introduces two other considerations: 1) is open audio feasible, and 2) can different types of signals (e.g., audio and graphics) be multiplexed (combined) over the same channel? As noted earlier, the use of open audio is a way of avoiding the need for voice switching. The use of multiplexing can reduce transmission costs.

We have not exhausted all the analog transmission possibilities, since other combinations of systems are possible. We have already discussed the combination of the MTS with dedicated circuits. Two others should be noted. First, MTS and one or more multipurpose private networks may be combined if the different parties use different types of systems when calling into a *meet-me* bridge (described in the next section). The quality will be heavily dependent on the quality of the non-MTS connection(s).

The other type of combination occurs when some individual connections to the bridge employ either 1) WATS lines or 2) the MTS for local ends, together with the dial-up trunk service of an other common carrier (OCC) (e.g., MCI, Sprint and others). Experts warn that a substantial diminution in quality can be expected, especially with the latter.

Digital transmission is infrequently used for audio conferencing at present. Two developments should, however, be noted. First, for some of the graphics adjuncts considered later there are interdependencies among resolution, transmission rate and transmission time. As a result, it may be

**Table 2.2   Differences Among Analog Audio Conferencing
Transmission Systems**

Factor	MTS	Dedicated Two-wire	Dedicated Four-wire	Multi-purpose Private Network
Transmission quality	Generally good but variable loss	Good	Very good	Not always acceptable
Accessibility	Virtually universal	Limited to one's selected sites	Limited to one's selected sites	Limited to particular network
Commercial bridging services available	Yes	No	No	No
Costs	No fixed costs, but no economies of scale	Unit costs may be low, if use is high	Unit costs may be low, if use high	Generally least costly
Open audio feasible	Generally, no	Generally, no	Yes	Sometimes
Multiplexing possible	No	No	Yes	Sometimes

necessary to use a channel with a higher information carrying capacity than the MTS in order to transmit a high resolution image in an acceptable time. Equipment is becoming available which relies on high-speed digital circuits in order to provide high resolution images.

Second, some of the video conferencing services described later in this chapter, such as that of Satellite Business Systems, can also be used in an enhanced audio conferencing mode (providing *freeze-frame* images). These rely upon digital transmission.

Over the long term, we anticipate a trend toward digital transmission of voice, as well as other signals, in audio conferencing. One reason is that it is easier to design good digital bridges than good analog bridges. (Indeed, there are bridges in use today which convert incoming analog signals to digital mode and outgoing signals to analog, so as to operate in digital mode.)

Conference Bridges

The purpose of a conference bridge is to interconnect the circuits from each location. Ideally, it should do so in such a way that it minimizes the *loss contrast* or matches the levels of the different circuits—i.e., the differences between their losses—and cuts out noise from those locations from which no one is speaking.

A number of bridges are of the *meet-me* variety, while others are operator-initiated. Some bridges can function in either mode. When meet-me bridges are used, a call is made from each participating location to the bridge. This spreads the start-up effort over all locations, saving considerable time when there are many sites involved. In this case, each participating site is responsible for its own toll charges.

In an operator-initiated audio conference, a call must be placed, in turn, from the bridge to each site. The calls are usually made by an operator, though there are desk-top bridges which can be operated by the conference originator when there are only a few participating locations (i.e., four or five). In this case, MTS toll charges are centralized and passed on to a single account.

As noted in our discussion of transmission, those who use private lines, whether dedicated to audio conferencing or multipurpose, must install their own bridges. Users of the MTS, on the other hand, have three options:

- To install their own bridge;
- To use the telephone company's conference operator;
- To use another bridging service.

The advantage of the telephone company's service is that it is simple and well known, and requires minimal effort on the part of conference originators. It has disadvantages, too. The operator cannot monitor transmission quality to correct technical problems. It is usually not easy to reenter the conference after being disconnected. And this service can be expensive, when compared to alternatives. The cost depends upon the tariff in force (see Chapter 7), but is generally a function of the full business day charge between the two farthest locations and the number of locations.

Currently, three companies (Connex, Darome and Kellogg) offer a meet-me conference call service; the number is likely to increase soon. With this service, participants call a preassigned number at a previously agreed time. Their identity is checked and then they are connected into a bridge. An operator is continuously available to monitor and correct technical problems. Supplemental services are available for an additional fee (e.g., planning and organizing major conferences, and providing for transmission of graphics).

With these services, the originator pays for the use of the bridging service—currently about $20 per port (connection to a bridge) per hour—and participating locations pay their own toll charges. The main advantages, compared to the telephone company's conference operator, are 1) a higher level of service, and 2) higher capacity in most cases, which is important if one wishes to involve a large number of sites (as when the audio component of a large one-way video, two-way audio teleconference is handled by the dial-up telephone network).

The other possibility is to install and operate your own bridge. Above a certain level of use, this may be less costly and can provide the flexibility of independence from outside bridging services. Relatively unsophisticated desk-top bridges with four ports cost as little as $200. More sophisticated bridges, with 40 or more ports, may cost in excess of $40,000, to which must be added installation charges (probably in the range of $1000 to $2000 a port) and monthly line costs.

Some electronic private branch exchanges (PBXs) incorporate a bridge, but usually the capacity is low—about four to six ports.

The location of the bridge in the network affects the ease with which it can minimize the loss contrast. *End-point bridging* occurs when the bridge is located at one of the end-points of the network. In *mid-point bridging,* it is located as closely as possible to the electrical center of the network. Most signals travel farther through the local ends of the transmission network in end-point bridging; it is these parts of their journeys that are most responsible for loss and other transmission disturbances (see Figure 2.3). For this reason mid-point bridging is superior in minimizing loss contrast. It is also better for controlling noise and echoes.

It is usually quite easy to locate a bridge close to the electrical center of a dedicated network. Only the telephone company, however, has access to the inner reaches of the MTS.

ENHANCEMENTS FOR AUDIO CONFERENCING

Between basic audio conferencing and two-way video conferencing is a range of services, known by various names: enhanced audio conferencing, audiographic teleconferencing and freeze-frame (or captured frame or slow-scan) teleconferencing. These are intermediate services in the sense of both technical capability and cost. They are not necessarily, however, intermediate in terms of effectiveness: in many cases they combine much of

Figure 2.3 Local Ends in End-point and Mid-point Bridging

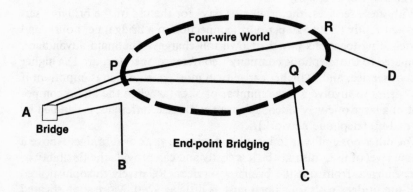

The bridge is at A. When D, for example, is listening to C, the signal travels from C to Q to P to A to P to R to D. There are four links outside the "four-wire world": CQ, PA, AP and RD. The "four-wire world" comprises Class 4 and higher order switching stations and the connections between them. The links outside it generally contribute much more to the degradation of a signal than the links inside it. In this example, there are four outside links.

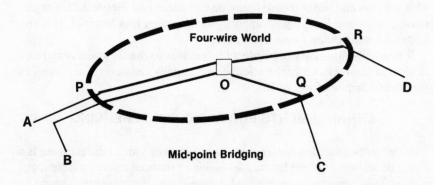

The bridge is at O, inside the "four-wire world." Now, when D listens to C, the signal travels from C to Q to O to R to D. There are only two links outside the "four-wire world": CQ and RD.

the flexibility and economy of audio conferencing with video conferencing's ability to convey important graphic information.

There are basically three kinds of enhancements:

- Remote control of projection equipment;
- Signalling to provide automatic speaker identification; and
- Transmission of real-time drawings or still television images.

Each will be reviewed in turn.

Facsimile transmission provides a fourth type of enhancement. Printed documents and diagrams can be transmitted before or during a teleconference. Diagrams can be printed immediately as transparencies for overhead projection. Facsimile transmission may operate over regular dial-up telephone lines or, if greater resolution and/or speed is required, it may operate via dedicated digital channels. Since this mode of telecommunication is quite well known, we will not discuss it further here.

Remotely Controlled Projectors

Equipment is now available which enables remote control of random access slide projectors and microfilm projectors at any or all sites. (The supplier is Revox Systems, Inc. The current price is between $700 and $1000 per site.)

To call up an image, the speaker uses a regular push-button telephone to key in the two-digit number assigned to the image (i.e., carousel slide position). The corresponding tones are carried as superimposed sounds in the audio transmission and serve to operate the projector, which then causes the appropriate image to be displayed. (The "bleeps" have no disruptive effect on the teleconference.)

The obvious disadvantage of this method is the need to send physical copies of slides or microfiche in advance. The major advantage is high resolution color graphics. The majority of today's real-time graphics telecommunication systems offer much lower resolution and are black-and-white. Another advantage is very low cost, possible because the signalling does not require an additional telephone channel.

Audio conferencing, supplemented by remotely controlled projectors, has been successfully used for presentations at scientific and professional conferences. It also has obvious potential for the purposes of education and training. It is much less likely to be of significance in business teleconferencing, given the lead time necessary to prepare materials. Remotely controlled projectors can also be used to supplement a video conference.

Automatic Identification of Speakers

Both the British and subsequently the French have developed equipment which automatically indicates who the speaker is at the remote location. Both systems are for use only over four-wire circuits between two conference rooms. Each is based on the use of a special conference table which provides a separate microphone for each of from four to six participants.

In the British approach (called the Remote Meeting Table), there are loudspeakers at dummy positions at either table, corresponding to each of the remote participants. Sound comes out of the loudspeaker allocated to the person who is talking at the far end, and a lamp on this loudspeaker is activated. In the French approach, there is only a visual signal: an LED (light emitting diode) display, adjacent to a list of names, indicates which person is speaking.

The French system is marketed in the U.S. by Optel Communications, Inc. (formerly FTC Systems). The price for a package including table, audio equipment, identification subsystem, basic acoustic treatment and a drawing device (electronic tablet) is currently around $25,000 per site.

From time to time experiments have been conducted in the U.S. with stereophonic sound, which would provide a directional indication of who is talking. This approach, however, has not yet taken root.

In various public demonstrations of enhanced audio conferencing, Bell Laboratories has shown how to provide the electronic equivalent of raising one's hand. Each participant is assigned an identification number. If one wishes to enter the discussion, he or she simply enters the number on a push-button telephone or a simple device (tone generator) that simulates its tones. The signal, superimposed over the regular audio, is picked up by equipment at the chairperson's location and displayed for his or her use. The display can hold several numbers in a queue; the chairperson can cancel the number after inviting the corresponding individual to speak.

When this equipment is used in conjunction with remotely controlled projectors, the identification numbers can correspond to photographs of the participants. Keying in one's number then causes one's photograph to be projected.

Writing and Drawing Devices

Various devices are available which transmit the X and Y coordinates of a moving pen, stylus or chalk via a regular telephone connection to distant sites where they are displayed on a monitor. These allow participants to draw diagrams or write to illustrate what they are saying in an audio conference. Alternatively, graphics can be recorded in advance (on an audio

The French system for automatic identification of speakers has an LED display which indicates who is speaking. Courtesy Optel Communications, Inc.

tape recorder) and called up for display and modification at appropriate points in a meeting. (For transmission over the telephone network, graphics signals must be made to correspond to sound; as a result, they can be recorded on a regular audio tape recorder.)

Some equipment, designed for other purposes, has been available for a long time. At one extreme are simple electromechanical devices, which hold a pen in a movable frame. When the pen at one location is held in contact with the paper, its movements cause the distant pen to replicate the movements of the first one. Output may be drawn on paper or on acetate film. If the output is drawn on acetate film, it may be projected onto a screen. Transmitters cost a little under $2000 each; receivers which incorporate a projector cost about $3500 each. The equipment, however, is limited and awkward.

At the other extreme is sophisticated electronic equipment, originally designed for use in broadcast television studios. It uses light pens to write on display screens.

In between there are various devices that can be used in enhanced audio conferencing. The best known is the Bell System's Gemini® 100 electronic blackboard, which was designed for use over a dial-up telephone line. The input terminal looks like a regular blackboard. Actually, the surface is a special film; behind it is the necessary circuitry to locate the pressure of chalk or an eraser. Output is displayed on regular black-and-white television monitors.

The electronic blackboard is particularly easy and natural to use. Excluding the monitor, there is at present a one-time installation charge of between $400 and $500 for each unit and a monthly rental cost of about the same amount.

Another approach is adopted in the French speaker-identification system described earlier. Here one may write or draw in two colors on a graphics tablet placed on the conference table. Output is displayed on a color television monitor at the remote site. Resolution is inferior to that of the electronic blackboard. However, as well as providing color capability, the French system incorporates a memory from which images, prepared in advance of an audio conference, may be called up.

Currently, the French system is available as part of the package described above. However, there are plans to make the electronic drawing system available by itself in the near future, presumably for operation over a regular dial-up connection.

A third method is likely to become available soon. It would be based on digitized graphics tablets which can be connected to personal computers and would provide a greater range of color and more flexible recall than the French system. The hardware at each site would comprise a suitable

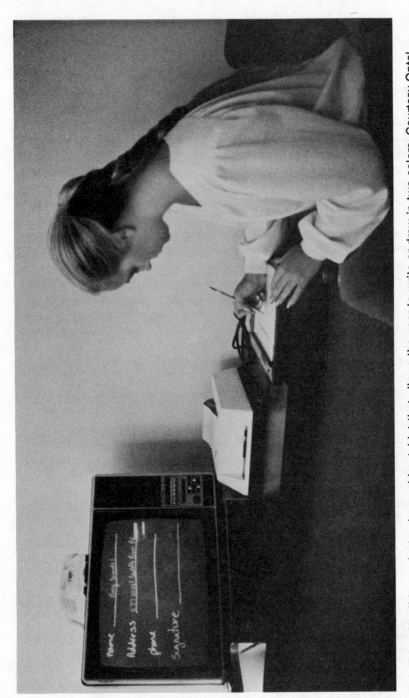

The Telewriter II is an electronic graphics tablet that allows the user to write or draw in two colors. Courtesy Optel Communications, Inc.

personal computer, a graphics tablet, a monitor and a modem,* for a total cost of about $4000. Such a capability has been demonstrated, but a software package is not yet available.

Slow-scan Television

Slow-scan television allows any *still* image which can be captured by a regular television camera to be transmitted over a dial-up telephone line. It can be used to show participants, graphics or objects.

Slow-scan television operates by trickling the information that comprises a television image down a narrowband channel. At the far end, it is assembled in a buffer (i.e., an information storage device, similar to a computer memory) and, when the transmission is complete, displayed on a television monitor. Transmission time is typically half a minute. There is, however, an interdependency between resolution, transmission time and the use of color, so transmission times vary considerably. Resolution is adequate for large lettering, but is not sufficient to display the whole of a regular 8½ x 11 inch typed page.

The term *freeze-frame* refers to a modification of slow-scan TV which deals with scenes in which there may be movement. A monitor is watched at the near end; when the image is judged appropriate, it is "frozen" in a local buffer. The frozen, hence static, image is then transmitted, via slow-scan.

Some freeze-frame systems operate with multiple cameras and a switch —e.g., one camera for graphics, two for people. Among the options available are 1) user control of the resolution-transmission time trade-off, and 2) an electronic pointer, which can be moved in real time around the displayed image.

There are three suppliers of off-the-shelf systems: Colorado Video, Robot Research and NEC America. The first two provide monochrome transceivers which, together with camera and monitor, range upward in price from about $6000 per site.† NEC America and Colorado Video offer color capability with units costing about $18,000 and up.

---

*A *modem* is a device which converts a signal from a computer into an analog form for transmission (in this case, over a telephone circuit), and converts the analog signal back into digital form for input into another computer or peripheral device. The cheaper modems for use with personal computers operate at a speed of 300 baud (one baud is one bit per second, a measure of information transmission speed). This application, however, would probably require 1200-baud modems.

†A *transceiver* comprises a transmitting and a receiving unit.

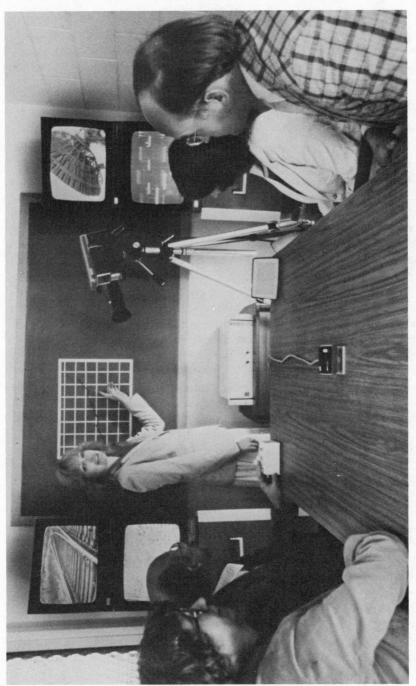

A slow-scan teleconference room with multiple memory picture storage capability. Courtesy Colorado Video, Inc.

Freeze-frame systems can be installed in conference rooms with preset cameras and push-button controls for the chairperson and/or facilitator at each site. Otherwise, an operator will probably be needed at each location. (It generally takes only a few minutes to learn how to operate a system satisfactorily.)

VIDEO CONFERENCING

As noted at the start of this chapter, there are two forms of video conferencing. Each raises rather different technological issues. Point-to-multipoint video conferencing, with one-way video and two-way audio, is akin to broadcasting. Technology and technique, and their interaction, are well understood. The addition of a two-way audio component need not seriously complicate the process: given the large number of people usually involved, only a low level of interaction is generally possible. When audio performance is inadequate, it is often a result of ignorance of the principles of good audio conferencing.

Two-way video conferencing for business meetings raises far more difficult issues. It is much more constrained: there are generally only two locations and up to only six participants at each. It replaces a highly interactive form of communication (face-to-face meetings). It is usually far more expensive on a per capita basis. And the state of the art is at a decidedly formative stage.

Because of its cost, transmission determines the design of a two-way video conferencing service. If distances are great, it may well be worth spending large sums on terminal equipment which will reduce the transmission capacity needed. Because the systems are expensive, they must be justified (so current thinking goes) on the basis of their uses for important purposes; this implies use by top management. All this suggests a "top of the line" approach, which encourages sophisticated design of video conferencing rooms. These generally end up costing between one-quarter and three-quarters of a million dollars. However, it is interesting to note that the current trend (with successfully installed video conferencing systems) is toward their use by middle management and by project groups.

When distances are shorter, a more relaxed attitude sometimes can be taken toward transmission costs and the needs of eventual users, who are less likely to be senior executives. This attitude seems to spill over into conference room design and the selection of terminal equipment. Less emphasis is placed on complicated camera equipment and sophisticated control systems. Although this phenomenon is seen in business teleconferencing, more extreme cases arise in the cable television arena where video conferencing is occasionally used for educational and public service purposes.*

The approaches adopted in two-way video conferencing over short distances (e.g., intra-urban) are highly varied and *ad hoc*. We cannot do them full justice in this chapter. Our treatment of video conferencing implicitly focuses more on long distance systems. We start with the subject of transmission, then move to the design of conference rooms and selection of equipment for them.

Video Signals and Their Transmission

To understand how a television camera converts a visual image into electronic form, imagine a grid superimposed over the image. (See Figure 2.4.) This divides the image into a number of picture elements, or *pixels*. Associated with each pixel is a certain level of light, its *luminance*, and a color, its *chrominance*. (The latter is irrelevant in the case of black-and-white television.) The camera scans row by row, along each row of pixels. As it scans, luminance and chrominance are converted into corresponding electric signals. When the electric signals are received, the reverse process takes place: each pixel on the screen is "given back" its luminance and chrominance.

Each *frame* (i.e., complete scan) comprises 525 lines (in the North American standard) and is scanned in one-thirtieth of a second. It is the number of horizontal scanning lines that determines the vertical detail, or *resolution*, in a picture. If much fewer than 30 complete frames were transmitted in a second, the displayed image would have the jerky appearance of an old silent movie. The number of complete frames transmitted in a second is known as the *frame rate*.

As it turns out, it is not good enough just to transmit a succession of complete frames, each in a thirtieth of a second—the decay of luminance in a pixel would be noticeable to the eye unless it was refreshed more frequently than this. Thus, each frame is divided into two *fields:* one of odd scanning lines (1,3,5,...525) and the other of the even lines (2,4,6,...524). Odd and even fields are transmitted alternately, each in one-sixtieth of a second, and no flickering is apparent.

This is how broadcast quality television is transmitted via analog channels. The channels, however, require a bandwidth of about 5 megahertz, which is expensive. Consequently, there is an economic incentive to make

---

*In the Irvine (CA) school district system, all the necessary equipment is on a trolley. The equipment can be plugged in at numerous locations in the buildings concerned. Regular surveillance cameras are used with available light. The cost of hardware at each location is less than $300.

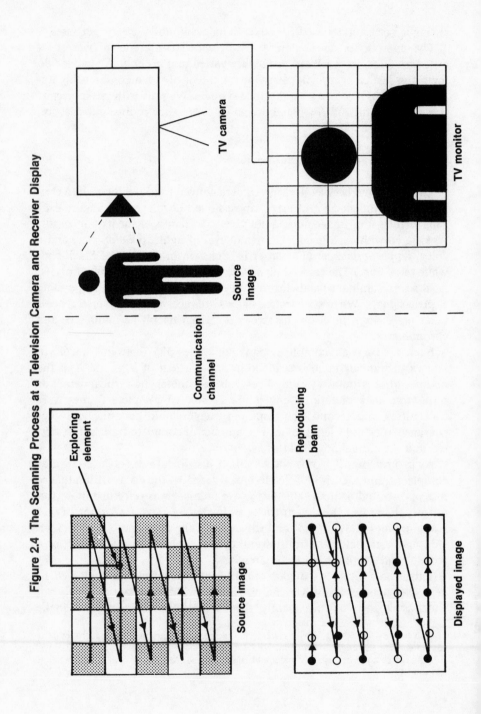

Figure 2.4 The Scanning Process at a Television Camera and Receiver Display

some sacrifice in picture quality, so as to reduce the amount of information per second which must be transmitted. This trade-off is acceptable because most video conferencing does not require broadcast quality transmission. For example, a video conferencing system generally does not have to cope with the rapid motion which occurs, say, in a sporting event. By exploiting such features, the sacrifice in picture quality is limited to a level which, it is hoped, is rarely perceptible.

Bandwidth Compression

Various techniques are used in combination for the compression of television signals.

Field elimination refers to the omission of one of the two fields, odd or even, which make up a complete frame.

Frame elimination reduces the number of full frames transmitted. For example, alternate frames may be transmitted and each received frame may be displayed twice. Obviously this and other techniques work better when there is relatively less motion.

Intraframe coding seeks out redundancy in the image captured by the camera, such as solid background, and transmits the corresponding information in a condensed form, rather than on a pixel-by-pixel basis.

Interframe coding relies on redundancy between successive frames. A coding algorithm identifies the difference between each frame and its predecessor. It is the difference which is then transmitted.

Codecs

The unit which receives the signal from a camera and applies these techniques is called a *codec* (coder-decoder). A similar codec at the far end translates the compressed signal back into a form which is recognized by the monitor as a regular television signal.

It is helpful to have an elementary understanding of the functions performed by codecs; more is involved than just compression and its reversal. The following description provides a rudimentary overview, without explaining the signal processing procedures. (See Figure 2.5.)

The process begins with the conversion of the electric signal originated by a video source (i.e., camera) from analog into digital form.* The signal is then demodulated. Essentially this means that the separate components of luminance and chrominance information are separated. The signals are then subjected to a combination of bandwidth reduction techniques.

---

*The first step temporarily increases the necessary bandwidth for carrying information. The subsequent compression stage more than compensates for this.

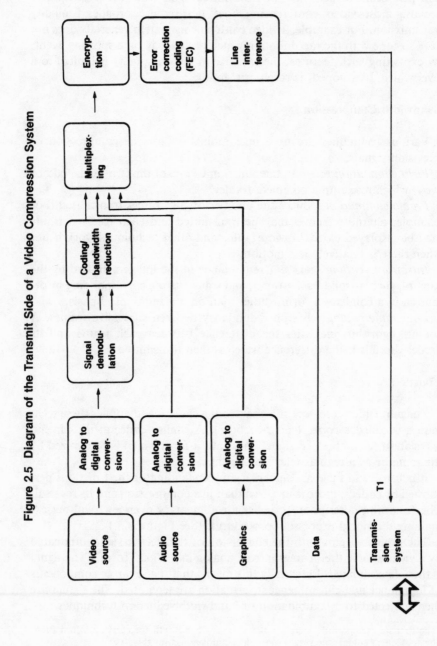

Figure 2.5 Diagram of the Transmit Side of a Video Compression System

At this stage, digital information from other sources can be added to the compressed video signal. The process of combining signals from different sources is referred to as *multiplexing*. These other sources may provide audio, graphics or data. (Data may relate to the operation of the control systems in the different rooms.)

If desired, the composite signal can then be encrypted for purposes of security. The final stage in the process involves additional coding for the purpose of controlling transmission errors.

When the composite compressed signal is received, essentially the same functions are carried out, but in the reverse order.

Video compression systems also incorporate a microprocessor-based diagnostic and control subsystem. Today's systems identify problems at the level of individual printed circuit boards. A user who encounters a problem then knows which board is defective and can replace it with a good one.

At present, video compression systems can be purchased from two companies: Compression Labs, Inc. (CLI) and Nippon Electric Co. of America (NEC). Each uses a different processing procedure, so the systems are not compatible.

Transmission Channels

In choosing between analog and digital (compressed) transmission, one is trading high fixed costs for codecs (at roughly $170,000 each) against high hourly costs for analog transmission. For point-to-point video conferencing, the higher the total time utilization (i.e., hours of use per week) of the channel, the likelier it is that digital transmission will be more economical. For point-to-multipoint video conferencing, analog channels are generally used today.

Analog Transmission

As noted earlier, analog signals can be transmitted via various media. Over intra-urban distances, a dedicated coaxial cable can be used (probably leased from the telephone company). Alternately, the necessary channels can be leased from a cable television company, although at present, this is unusual. Over intermediate distances (e.g., a 20-mile connection from downtown to a suburban location), the use of microwave transmission is quite likely. In predominantly rural areas, it may be better to install and operate a private microwave network. In urban areas, however, congestion of the airwaves may make it impossible to obtain a license. Even so, it may still be possible to lease the necessary transmission from the operator of a microwave system.

When distances are long (more than 200 to 300 miles), it is likely that a communications satellite will be used. The economic advantages of a satellite are twofold. First, cost is independent of distance for the long-haul portion of the connection. Second, it costs no more to transmit to many satellite ground stations than to one. This is what makes point-to-multipoint video conferencing economically viable.

It is important to remember that the short-haul costs of local ends may be quite high. It all depends on the location of the video conferencing site in relation to the nearest suitable satellite ground stations. For point-to-point video conferencing, at present, this is largely a matter of luck, which is why it is difficult to generalize about the actual transmission costs of end-to-end service. For point-to-multipoint video conferencing, on the other hand, one generally can select one's sites so as to eliminate most of the costs of local ends. Television receive-only earth stations (TVROs) add to locational flexibility, but currently they are expensive.

Digital Transmission

Digital signals are also carried via different media. The key point about digital transmission, however, is that it comes in standard packages defined according to the data rates of the channels concerned. One may purchase T1, 2T or T2 channels which correspond, respectively, to data rates of about 1.544, 3.0 and 6.0 megabits per second.* The data rate required depends on the codecs being used. Thus, the Bell System's Picturephone® Meeting Service, currently based on NEC codecs, uses one or two T1 carriers. (The NEC codec can be switched between 1.5 and 3.0 megabits. Bell includes a third T1 carrier for back-up purposes.) The CLI codec, however, is used with only one T1 carrier.

One can lease an end-to-end T1 carrier from the Bell System, though this service is not available nationwide. Satellite Business Systems provides digital transmission including T1 carriers for its customers as well, but customers must also pay for the necessary ground stations. Otherwise, T1 carriers can be leased from operators of communications satellites or middlemen; however separate arrangements must be made (probably with the telephone company) for transmission to and from the corresponding ground stations.

---

*The terms T1, T2 and T3 were coined by the Bell System to name channels with different data rates, and have become industry standards. The term 2T is used by some engineers to refer to the equivalent of two T1 channels. A single satellite transponder has capacity for about 20 duplexed T1 carriers, but only one analog television signal. This is why compressed digital transmission is economical.

A teleconference conducted via the SBS system in a room constructed for demonstration and development purposes.
Courtesy Satellite Business Systems, Inc.

A teleconference conducted via the Bell System's Picturephone® Meeting Service. The Bell System plans to link 42 cities with its video conferencing network by the end of 1983. Courtesy AT&T.

Technological and commercial innovation is proceeding at a rapid rate in the field of digital transmission. Additional options will become available during the next few years. Selection of the best option may then become quite complicated.

Of particular interest, in this context, are digital termination systems (DTS), which the FCC has recently authorized for major urban areas. These will employ a specially reserved frequency range for digital transmission via microwave between satellite ground stations and office buildings within a radius of a few miles. The concept calls for one antenna for each of three 120° or four 90° sectors centered on the DTS station. Time division multiplexing (TDM) will be used to enable systems to serve several customers at the same time. (In time division multiplexing, a very short time interval is devoted to burst transmission for one customer, the next interval for another, and so on; when the cycle is completed, it starts over again. The burst duration is a function of how much information there is to transmit and how many stations are included.) Engineering trials have been completed successfully, but commercial services are not yet available.

DTS services will provide T1 carriers, as well as lower speed channels. It is possible that, in many cases, they will be more economical than other options in providing local ends for bandwidth-compressed video conferencing services.

Room Selection and Support Systems

Conference Rooms

In point-to-multipoint, one-way video, two-way audio conferencing, conference rooms are usually regular meeting rooms in hotels or conference centers. Room design is not a major issue, though appropriate lighting, proper placement of monitors and the use of large-screen projection systems when numbers so require are not to be treated casually. Also, it is important to select a room which is acoustically desirable and, if audio interaction is required, to consider the other factors identified earlier in the audio conferencing section.

In the case of point-to-point video conferencing the situation is quite different. Each room will probably cost between $250,000 and $750,000. It is important to have proper design. Since the theory of designing a video conference room is still very much at a formative stage, today's design teams often draw upon a variety of skills relating to audio and video technologies, transmission systems, industrial design, ergonomics* and psychology.

---

*Ergonomics, or human factors engineering, is the science of designing machines, operations and work environments that best meet the needs of the workers involved with them.

The first step is to select a suitable room at each site. Criteria are obvious: it should be large enough to accommodate comfortably the intended number of participants and the equipment; it should be located close to the prospective users; and it should be in a quiet area. But don't underestimate the difficulty of securing prime space, or any other space for that matter, for a video conferencing room.

Lighting

Lighting will have to be changed to produce a quantity and quality of light which is technically appropriate for the types of cameras used. (Color television requires more light than black-and-white television. As a result, it creates more heat.) Proper lighting is necessary for perception of depth, accurate rendition of color, to avoid misperception of shape and, more generally, to create the desired atmosphere—and a pleasant picture.

A four-point lighting scheme may be used. This consists of one set of directional front lights (*key lights*); *fill lights* to offset the harshness and cancel the shadows created by the key lights; *backlighting* to highlight hair and clothing and give a three-dimensional impression; and *background lighting* for the back and sides of the room to separate meeting participants from the background and, as necessary, for a podium or blackboard.

In television studios one has the luxury of being able to reset the lights each time they are used. Not so in video conferencing. Accordingly, the lighting must be arranged to provide the best possible compromise, taking into account future variations in group size, heights of participants and their facial coloring.

Problems can be caused by reflection from a conference table or from other objects. The former can be controlled by having a nonreflective surface on the table. Alternatively, you may be able to incorporate the effect as a deliberate part of the lighting scheme.

HVAC Requirements

Heating, ventilation and air-conditioning (HVAC) requirements must also be considered. They will be affected by the equipment used, especially by the heat created by lighting, and also by local fire codes. It may be necessary to reduce noise transmitted through HVAC systems by installing "noise reducers" at the duct heads.

Because of time zone differences, the video conferencing system may be used outside regular office hours. Also, to avoid continual powering down of the compression and control systems (which is not recommended), it is probable that much of the equipment will be powered up between conferences. Consequently, there should be thermostatic control in the room and any necessary adjustments should be made to the timing of the building's HVAC systems.

A final consideration is electrical power. It is likely that additional power will be required to support all the equipment used. Contact breakers should probably be installed, and line conditioning equipment to protect against power surges and brown-outs is a wise investment.

Room Design and Terminal Equipment

Current wisdom emphasizes the importance of simulating an in-person meeting as closely as possible. The idea of "virtual space" has grown up: one tries to create the illusion that the distant parties are at opposite ends of the table. In other words, the video monitors should be windows through which participants see others facing them across the table. (See Figure 2.6.)

The Conference Table

Four shapes of conference table are commonly used: rectangular, crescent-shaped, octagonal and trapezoidal. Seating arrangements are illustrated in Figure 2.7.

The weakness of the rectangular table is that it is difficult to see others seated at the table. A self-view monitor alleviates this problem, but it introduces another problem: it can make participants, especially newcomers, self-conscious to see themselves on camera.

The crescent shape largely overcomes the problem associated with the rectangular table and avoids the need for a self-view monitor. It is important that the curve be such that differences in the perceived distance of participants from the "window" are not too distorted.

The octagonal table offers a compromise. It offers better intra-room eye contact than the rectangular table and it positions participants in a closer optical plane than the crescent.

The trapezoidal table can seat participants at three of its four sides. Most participants in a room have a clear view of one another. However, only the person at the head of the table—usually there is only one—faces the camera. The others must turn their heads to see the monitor.

The Audio System

The same principles apply to the audio system as in audio conferencing. The audio component of video conferencing has too often been disap-

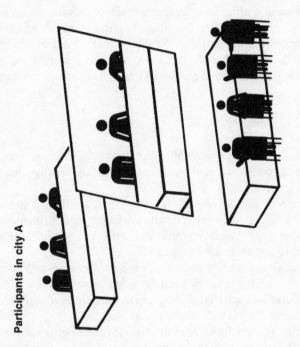

Figure 2.6 The Video Display "Window"

Participants in city A

Participants in city B

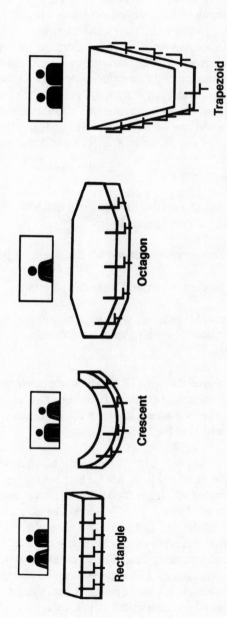

Figure 2.7 Various Table Shapes Used in Teleconferencing

pointing in the past, presumably because too much attention has been diverted to the more glamorous video component. There is no excuse for this mistake to be repeated in the future.

In a permanently installed conference room, loudspeakers will probably be located in the wall, facing the participants, or in the ceiling. Lavaliere microphones can be clipped to clothing or on a cord around people's necks, or microphones can be in stands on the table. Pressure zone microphones (recently introduced by Crown International Inc.) can also be used. These can pick up sound generated almost anywhere in a room. Another possibility is to use Bell's Quorum® linear array microphones. There should not be any unsightly cords in view and, if microphones are visible, they should be nonreflective to avoid glare.

The Video System

In transmitting visual images of participants, four approaches may be distinguished (see Figure 2.8):

1) A single preset camera which covers all participants.

2) A single camera and human operator.

3) Multiple cameras, with switching to determine which of them is to transmit. Switching may be manual, automatic or a combination of both.

4) Multiple cameras, all of which are active and which together cover all participants. Images may be automatically combined in a *split screen* or the combined image may be divided among several monitors placed side by side.

The first approach is the simplest and cheapest, but it does not allow a close-up view of the person who is speaking. As a rule of thumb, you can tell relatively little from facial expressions if more than two people are on camera at the same time. The second option allows for both close-up and ensemble shots. It is the only mode that is anticipatory: the person operating the camera can transmit a picture of a person who wishes to get into the discussion; automatic switching is only reactive. (In theory, manual switching by a chairperson can be anticipatory; in practice, it rarely is.) However, human operators represent a high continuing cost and may provoke concern about confidentiality.

Usually, the third approach is implemented by transmitting from the camera trained on the position at which someone is speaking. Cameras may be set to transmit one-person or, more likely, two- or three-person shots. Automatic switching is triggered by a microphone picking up a voice (or, unfortunately, a cough). After a short interval of silence at a location, transmission reverts to an overview camera which covers the whole group.

A variation on the third option is to employ a split screen. One portion of the screen can provide a group shot, the other a close-up of the person speaking.

The fourth approach represents two different modes. When images are combined in a split screen, the purpose is to devote more of the screen to participants, less to background. This allows for images of each member of the group that are larger than those provided in the first option. However, when the overall image is divided among adjacent monitors, the purpose is to provide simultaneous close-up views of all participants. This is sometimes termed the "continual presence" mode. Naturally, it requires more transmission capacity or a reduction in picture quality.

Which approach to select is in part a matter of cost. It is also a matter of one's philosophy as to the importance of being able to see all the distant participants at any time, relative to the importance of having a good view of the facial expression of the person who is speaking. Even a good close-up, however, will not provide real eye contact.

The state of the art is immature. There is no agreement as to how the various approaches differ in their effectiveness and acceptability. Experimentation will undoubtedly continue.

Monitors should be located as close as possible to cameras to provide as good a simulation of eye contact as possible. Their size should be in the range which, at the high end, would yield a slightly larger than life-size image, and, at the low end, would be what users are accustomed to seeing on their home television sets (i.e., an at least 17-inch screen).

The number of monitors varies from one installation to another. The "continual presence" approach may require three or four monitors for the remote participants, while the other options require only one. In addition, you might have a monitor on which to preview graphics, prior to their transmission, and a self-view monitor. These two functions can be combined in one monitor, which can be used for either purpose. (There is no agreement on whether a self-view monitor is a good or a bad idea.) You can also have a separate monitor for the display of incoming graphics.

In selecting cameras and monitors, it is important to keep in mind any limitations in quality imposed by the transmission system. Since bandwidth-compressed systems do not provide broadcast quality video, it is unnecessary to buy broadcast quality cameras or monitors for them. JVC, Sony, Ikegami and others sell very good cameras in the $10,000 to

Figure 2.8 Methods of Displaying the Video Image

**Full screen display of
all participants—no
camera switching**

**Multi-screen display—
no camera switching,
"continuous presence"**

**Split screen—bottom
third shows all par-
ticipants; top displays
speakers (active camera
switching is employed
for top screen)**

Figure 2.8 Methods of Displaying the Video Image (cont.)

**Split screen display of
participants—no
camera switching**

**Single screen displaying
speaker(s) only—active
camera switching is
employed**

$30,000 price range; these cameras are highly suitable for use with compressed transmission.

The Graphics System

Graphics and printed materials are widely used in business meetings. This was illustrated by a recent survey of more than a thousand meetings conducted within a major U.S. corporation. It revealed the following pattern of use:

Material	Percentage of Meetings
Typed pages	60%
Transparencies	56%
Blackboard/whiteboard	28%
Sketches	26%
Flip charts	24%
Computer output	17%
Slides	15%

The most common method of transmitting graphics involves the use of a camera with zoom lens mounted in the ceiling over a designated area of the conference table, and/or a camera directed toward a wall-mounted unit which doubles as a blackboard and a screen for slide projection. When this camera is in use, it generally takes over the transmission channel otherwise used for the outgoing image of the participants in the room.

If hard copy is required there are two choices. A suitable printer can be coupled to the incoming graphics monitor, or copies can be transmitted by means of facsimile equipment. Group 3 facsimile systems* are generally used, since they require less than a minute to produce copies.

High resolution graphics transmission systems have recently become available as teleconferencing adjuncts. They offer black-and-white images of twice the resolution of regular television (more than a thousand lines, compared to 525). However, they are expensive and they require a digital channel of 488 kilobits per second, which means that they cannot share a T1 carrier with the compressed video signal used for the participants.

The Control System

The greater the number of video sources available at a site, the greater the need for a sophisticated control system. A comprehensive control system may need to cover:

- Switching among "people" cameras (manual and automatic);
- Switching between "people" and graphics or slide cameras;
- Pan and tilt of cameras;
- Zoom and focus of cameras;
- Play/record for video tape recorder;
- Allocation of monitors to outgoing signals (preview or self-view).

A fully manual approach might require close to 20 push-button controls and one or two joysticks for the cameras. This level of complexity would probably overwhelm a chairperson or his/her assistant. Complexity at the person-machine interface is reduced by one or a combination of three

---

*In an effort to promote compatibility of facsimile machines, the CCITT (International Telegraph and Telephone Consultative Committee) has developed three categories of facsimile machines:
> Group 1 devices use FM analog modulation machines and transmit a page of information in four to six minutes;
> Group 2 systems use AM analog modulation and transmit a page in two to three minutes;
> Group 3 machines employ digital compression techniques and transmit a page in one minute or less.

methods. First, options may be eliminated—for example, by the use of preset cameras. Second, switching among "people" cameras is usually automated, being driven by the activation of microphones. (A manual override may be provided.) Third, a microcomputer may be used. Only the control options that are relevant at any time are displayed; input is by means of a touch-sensitive screen.

CONCLUSIONS

In this chapter we have described the technological components available for audio and video teleconferencing, and have explained the principles underlying their combination into effective systems. We have distinguished four main types of teleconferencing—audio, enhanced audio, point-to-point (two-way) video and point-to-multipoint (one-way) video—and shown that there are many different systems of each type. The extent of this variety is often underestimated, especially by newcomers to the field.

Teleconferencing can fail because of technological mistakes, but it can also fail because it is based on an inappropriate concept of need or because it is poorly implemented. Parts of the field are still relatively immature, notably two-way video conferencing, and offer few, if any, proven models of needs assessment, technical design and implementation.

The four types of teleconferencing that are the subject of this book fit different kinds of needs. Only in relation to specified needs does it make sense to compare them in terms of effectiveness. It is as meaningless to consider audio conferencing to be a poor man's video conference as it is to consider an automobile a poor man's truck.

Unfortunately, the concept of need is elusive. Needs assessment is at best an art, certainly not a science. A primary reason is that domestic teleconferencing probably has less to offer as a substitute for travel than as *a substitute for communication that does not take place but ideally should.* It permits electronic meetings when conventional meetings are ruled out by cost, distance or scheduling problems. And it allows a more appropriate mix of participants, since the marginal cost of adding people to a teleconference is much less than the marginal cost of taking them along on a business trip. A sound approach to needs assessment, then, would be to study the communication that does not take place (but in some sense should), in addition to examining the meetings that do take place—in all, a very difficult proposition.

Likewise, implementation is no easy matter, especially since it involves embedding a new technology in an existing organization. In addition to the myriad problems concerning vendors, logistics and technical matters, there is the more difficult issue of overcoming individual and collective resistance to change.

Improved methods of needs assessment and implementation are as important to the development of teleconferencing as the new and improved technologies that are bound to emerge. Regarding the future of teleconferencing technology, the following developments seem likely to occur over the next five to 10 years:

- "Fixes" for emerging problems of incompatibility of terminals;
- Portable terminals for full-motion video;
- Improvements aimed at reducing the costs of video transmission;
- Overcoming the two-site constraint in business video conferencing;
- The emergence of integrated, modularly designed systems;
- New options for solving the "last-mile" problem in video conferencing.

Each of these topics will be described briefly.

Future Developments

Currently, different manufacturers' freeze-frame terminals are incompatible, as are codecs. The problem is likely to get worse with the growth of international teleconferencing. Unless and until there is standardization, it is plausible to expect microprocessors to be developed to handle some of the problems of translation from one standard to another. However, other problems will be too expensive to tackle in this way.

It is somewhat surprising that until very recently, relatively little thought had been given to making the more expensive terminals portable, so that they could be easily wheeled from one conference room to another. More ambitious, but still sensible, would be portable units which could be easily and safely shipped from one site to another. Potential advantages of portable systems include higher utilization of expensive equipment; flexibility in the allocation of conference rooms, hence greater convenience for users; and the ability to accommodate sites which have only occasional, but significant, needs. There would be disadvantages, too, principally arising from the use of the equipment in less desirable acoustic and visual environments, but the drop in quality might often be a price worth paying.

For codecs, the state of the art is developing rapidly; the bandwidth required for video transmission continues to fall. The limit to which it can ultimately drop depends on the acceptability of the received video images and is still the subject of considerable uncertainty.

We noted earlier in the chapter that, as yet, there is no generally accepted means of allowing full two-way video conferencing among three or more sites. The *ad hoc* solutions which have been adopted in the fields of telemedicine and education and in a few two-way cable TV systems would generally be too expensive in terms of transmission and/or labor if used in a business environment. The problem has been receiving attention, however, and is likely to be solved.

Ideally, an integrated system would be available that could provide any of the audio or video conferencing capabilities according to the modules used. One would be able to start with a less powerful system and add capabilities by adding modules. Currently, such a system is not available, but steps have been taken in this direction and some larger users are designing their own modular systems.

We can also expect progress on another aspect of integration: the integration of teleconferencing systems with office automation systems. Rather surprisingly, these have been two separate fields of endeavor until now. In principal, however, it would make sense to incorporate teleconferencing features into an office work station, and such systems are under development. Their advent will depend more on the market than on technological considerations. As long as potential customers perceive teleconferencing as a tool for management, and office automation as a tool for secretaries, such systems will be hard to sell. This, however, no longer represents the best thinking in the office automation field; customers' perceptions *are* changing.

The final development, the "last-mile" costs of video transmission, was discussed earlier in the chapter. We noted, in particular, that digital termination systems are soon to appear. Other solutions are likely to emerge involving cable television, small portable earth stations and local area networks.

Though seemingly less dramatic than the video-related developments, there will be steady progress in improving the quality of audio. Its impact could be considerable. Users will find teleconferencing less of a strain; voice recognition will be improved; it will be easier to interrupt and be interrupted; users will be able to hear without working as hard to listen; participants will be less constrained by the position of microphones.

Valuable as the technological progress of teleconferencing will be, however, it is important to realize that today's equipment already allows us to solve many more communications problems than we are currently solving. As the following chapters reveal, there is no need to wait for tomorrow's technology.

3

Issues and Problems in Teleconferencing

by Eugene Marlow

INTRODUCTION

A Stone in the Pool

As manager, Corporate Video/AV Communications for Union Carbide Corp. (UCC), I was working quietly at my desk in early January 1980, when I looked up and saw one of my clients staring at me from the doorway. I invited him in, and asked what he had on his mind. "We want to do a teleconference," he said.

My head flooded with excitement. One of our clients was considering the use of teleconferencing, and full motion! He was asking for details and facts. There was a lot of work to do.

My client had thrown a small stone into a pool of unknowns. Nine months later, the first ripple took shape.

In October, top Carbon Products Division managers and the president walked into the reception area of AT&T's Picturephone® Meeting Service (PMS) room in New York City. Several other managers from the division's research and development center were escorted into a PMS room in Cleveland, OH. The full-motion video conference was scheduled to start promptly at 10:30 a.m. (For 15 minutes we sweated out a technical problem in the audio portion of the system. Moments before the executives' arrival the problem was solved.)

To provide a record of the company's first use of full-motion video conferencing the producers had arranged for video tape coverage at each location. We also planned to interview each participant, individually, in each location immediately after the teleconference ended.

As the New York managers were shown to their seats, on the monitor we could watch the Ohio managers going through the same procedure. I thought, "This took nine months: meetings with the client, planning sessions with AT&T representatives, costs to figure, procedures to iron out, briefings for the executives. How will they react to this new environment? None of them has ever done a video conference before. Will they adapt? Will they freeze up? What will the division president think? What if the system breaks down? Will the audio hold up?"

A few minutes later an AT&T representative introduced herself and explained the system, the controls for the graphics camera, the voice-activated cameras. The participants at both locations heard and saw the orientation simultaneously. It was almost 10:30 a.m. The AT&T representative gave over control of the system to the division president and sat down.

The first few minutes were a little awkward—people didn't quite know how loud or soft to speak, or how to move; speech was a little stilted; thoughts came out too abbreviated. However, after no more than five minutes, it became clear—very clear—that a meeting was taking place, even though it was electronic. All awkwardness and self-consciousness vanished. The technology became transparent; it was simply a vehicle. The metamorphosis from "what is this?" to "let's get down to business" had taken less than five minutes!

I was amazed. I thought back to that first conversation with the division's communications manager.

"Why do you want to do a teleconference?" I had asked. The manager answered, "It will save us money, travel costs, and wear and tear on our executives, who travel back and forth from headquarters to our research and development center. The travel costs are easy to compute. The question is how much will it cost to set up our own in-house system? And how much will it cost to run it?"

It was a few minutes after 11 a.m. The meeting was proceeding at a lively clip. The exchanges were crisp and to the point. The division president, who originally had agreed merely to observe, was participating fully. The repartee was not only informative, but occasionally humorous. Graphics were used and valuable information was being exchanged.

At 11:30 a.m. the meeting concluded. The system had performed. At each location participants were interviewed for a few minutes.

In New York there was a lively conversation at lunch: about systems cost, maintenance costs, security, how often the system would be used, travel cost savings, wear and tear on executives. Most of all, the participants underlined their strong feelings that a real meeting had taken place.

Several days later, my staff viewed an edited version of the meeting, which included portions from each of the dozen or so interviews which were videotaped right after the teleconference. The consensus: 1) a real meeting had taken place; 2) the technology made everybody stick to the agenda; and 3) the simpler the system the better.

A year later (October 1981) at the Annual Union Carbide Video Conference, virtually the entire first day was devoted to teleconferencing of various kinds: audio only, enhanced audio and full-motion video conferencing. Speakers included representatives from Connex International, AT&T and Aetna Life and Casualty (which has installed a highly successful full-motion video conferencing system—see the case study in Chapter 8). Also present at the meeting were representatives of UCC's Telecommunications Department which was launching enhanced audio (slow-scan) conferencing services to its clients.

As of this writing, UCC's Video Communications Department and Telecommunications Department are working in concert to develop teleconferencing in the corporation. The potential is for a nation-wide teleconferencing system involving voice, data and motion video among a multitude of locations.

CURRENT STATUS OF TELECONFERENCING

In two and a half years UCC moved from a trial of a new technology to a more aggressive stance. Of course, UCC is only one of many organizations that have used teleconferencing as a "bridge across the waters" of a communications problem:

- The American Aviation Association: a portion of a national convention included a satellite conference between the Las Vegas delegates and government officials in Washington, DC, to discuss the environmental impact of crop dusting.

- Lanier Business Products: used a satellite to beam a national sales meeting to 12 different cities.

- The Picker Corp.: held a satellite meeting for 450 sales people in 30 Holiday Inns near their homes at a cost of $85,000.

- American Hospital Association: at its 1979 convention, transmitted by satellite 18 instructional programs to an estimated 17,300 hospital personnel in 425 hospitals, nation-wide.

- American Soybean Association: to highlight its 1979 conference in Atlanta, GA, more than 1500 convention attendees watched and heard soybean economists in Atlanta, Japan, England and Brazil discuss market outlets for the crop over a live satellite hookup.

- The Ford Motor Co.: used teleconferencing for a national sales meeting of 14,000 Ford dealers and sales representatives who gathered at 38 locations to participate in a program beamed at them from Detroit by satellite.

- TRW: held a seminar via teleconference in about 30 cities designed to attract its credit report customers.

- Texas Instruments: used satellites to telecast its 1981 Annual Stockholders Meeting from its Dallas headquarters to 22 locations across the United States.

- Merrill Lynch: used video conferencing for a two-hour program on the implications of the Economic Recovery Act of 1981. The company leased time on the Westar III satellite and beamed the program to convention halls and hotels in 30 cities.

There are countless more examples of the use of contemporary telecommunications technology for audio only and enhanced audio conferencing by companies such as IBM, Massachusetts Mutual Life Insurance, Datapoint Corp. and American Airlines (see Chapter 8).

THE ECONOMIC ISSUES IN TELECONFERENCING

What steers a corporation toward teleconferencing? The major impetus for the 1980 Union Carbide teleconference (apart from curiosity) was *productivity*, i.e., the cost of managers' time and travel and other per diem expenses, incurred in face-to-face meetings between the company's headquarters and the division's R&D facility. In the case of Aetna Life and Casualty, its in-house system links two facilities which are only nine miles apart. One of the major reasons for the operation is the high level of interaction among personnel at both locations and the high, cumulative cost of staff travel between the two locations (prior to the existence of the system).

Thus, the economic issue has two aspects: the cost of lost managerial time spent travelling among various locations, and the expense of transporting the managers among those locations. Let's look at the second aspect. When a corporation contemplates the use of teleconferencing it must consider the *present* and *future* costs of transportation. To determine if this "cost" is even relevant, an organization must perform a detailed analysis of personnel travel patterns. Some of the questions to ask include: Who travels from where to where? How frequently? What is the distance? What is the mode of travel? What is the reason for the trip? How long a meeting takes place once the manager gets there? Could a teleconference substitute for all, some or a small portion of the travelling?

Answers to these questions must be blended together with an analysis of the potential cost of a teleconferencing system. When these two analyses are combined, the answer may or may not support the development of an in-house teleconferencing system, or the use of an outside system.

The cost-effectiveness of teleconferencing does not rest solely with whether it saves an organization money; there are other less tangible benefits. Teleconferencing should enable managers to be more productive. Martin Elton, professor of communications at New York University and one of the authors of this book, feels that teleconferencing provides new opportunities for communications. He states:

> It is important to see teleconferencing as an alternative to noncommunication, not just as an alternative to an in-person meeting. It is also helpful to consider the role of teleconferencing in the mix of communication media that will be used through time.[1]

Elton also points out that an ability to communicate with persons at a distance may just lead to more in-person communication with persons at a distance; moreover, a teleconference can be used ". . . to plan agendas for expensive in-person meetings and to control the follow-up of action points after them."[2]

In my experience and that of others, teleconferencing (whatever the form) is not a replacement for travel; it is another way of communicating, another medium in the business communications environment.

Economics: The Supply Side

So far, I have dealt with issues relating to the "demand" side of teleconferencing's potential: reducing travel costs, increasing productivity, enhancing management communications, and so on. However, the relative cost-effectiveness of teleconferencing (and here I mean data, voice and video teleconferencing) will be affected by the supply of technology to satisfy such demand.

From one vantage point the cost of teleconferencing may come down for no other reason than the proliferation of systems. According to a 1981 report prepared for the Corporation for Public Broadcasting by Browne, Bortz and Coddington, there will be approximately 375 to 500 basic video conferences in 1985, as compared to approximately 50 held during 1981.[3]

More broadly speaking, an estimate by United States Satellite Systems, Inc. projects enormous growth in demand for teleconferencing as reflected in the growth of satellite transponders. According to its analysis, between 1985 and 1995 there will be a tenfold increase in the number of satellite transponders needed for video conferencing (from 20 to 200). All told, the number of transponders for data, voice, video conferencing and video (e.g., cable) will almost quadruple, increasing from 433 to 1550 between 1985 and 1995.[4]

Regulatory Issues in Teleconferencing

An implicit but extremely important issue that grows out of these projected supply increases is the impact the growth of telecommunications will have on government regulation. Here the issue is: Given the growth of telecommunications and the ways these new technologies are affecting the marketplace, how will government regulate this industry? And, therefore, affect competition and pricing?*

The changing regulatory environment makes predictions about future costs difficult. Which way might things go? Some argue that those corporations who do not buy into teleconferencing at today's prices will suffer substantially increased prices in the years ahead. If such estimates are correct, even though capacity will increase, demand will outstrip capacity, resulting in higher prices—this despite the probability of improved and lower cost hardware. On the other hand, there might be periods when supply exceeds demand (which could happen in the next few years as more satellites are launched) with the result that prices might stabilize or even drop.

---

*This question was addressed in a comprehensive 1981 report by the Subcommittee on Telecommunications, Consumer Protection, and Finance of the Committee on Energy and Commerce (House of Representatives), entitled: *Telecommunications in Transition: The Status of Competition in the Telecommunications Industry.* The report recognized that while telecommunications has been dominated by a few organizations, the changing nature of telecommunications technology (e.g., satellites) has contributed to a changing marketplace and hence competitive environment.

In either case, organizations must be careful not to presume that increased capacity together with demand will provide a cost-effective argument for the use of teleconferencing technology, whether *ad hoc* or in-house. Future demand must be defined and measured against future capacity and technological capability. The waters here are somewhat murky and should be entered cautiously. At the very least, managers should thoroughly research alternate means of teleconferencing before committing dollars to the project.

TECHNOLOGY AND COMMUNICATIONS PATTERNS

These economic incentives and problems notwithstanding, the evidence points to a widespread acceptance of teleconferencing because of:

- The higher costs of travel;

- The availability of additional satellite capacity and other long haul transmission facilities in this decade;

- The development of low-cost hardware;

- Numerous existing earth stations combined with transportable earth stations;

- The move by hotel, motel and conference facilities to offset reduced travel by offering teleconferencing services throughout the United States.

In what kind of communications environment, then, will organizational managers find themselves in the years ahead? Teleconferencing use might follow various scenarios. For example, some corporations will perceive that their needs are not frequent enough to warrant an in-house system. They will avail themselves of outside systems (e.g., AT&T's PMS® system) and turn to outside suppliers to set up turnkey systems on an as-needed basis. Other organizations will develop modest in-house systems with enhanced audio capabilities. Still others will have full-blown video teleconferencing systems with domestic and international hookup capabilities. Systems will have sophisticated terminal devices for the instant exchange of documents and information. They will be highly user friendly. In effect, teleconferencing systems could become as common as the telephone, and the use of the systems could be as easy and practical as picking up the conventional telephone is today.

Teleconferencing rooms will be, in a manner of speaking, akin to large telephone booths used by several persons at a time. Special acoustics for high quality audio will be built into the rooms. Specially shaped desks, perhaps even larger, high definition television screens, properly designed lighting and graphics camera controls will be available. No special engineer or technician will be needed, except when the system breaks down. System reliability will be high and the cost per use will be reasonable.

More than likely, organizations will go through a phased development of teleconferencing. First, they will use an *ad hoc* system; then more complex audio conferencing systems beyond the telephone conference; then slow-scan or freeze frame equipment; then, as the demand grows, full-motion video (in color) will be installed.

Resistance to Change

The transition will not always be smooth between the realization of these scenarios and the current status of teleconferencing, however. As two authors point out:

> Some of these services, and the equipment which provide them, will have an aspect of space age glamor. This will add to their appeal for the future-oriented, novelty-seeking segment of society. It will not, however, smooth the adaption of a slow-speed, paper-trained people to the new instant information culture. There will be resistance to these changes on a larger scale than ever encountered before. Even where resistance has faded, many "progress-produced" strains will threaten the peace and productivity of human society.[5]

Thus, while many advantages have been ascribed to teleconferencing, there are other "perceived" disadvantages. Principal among them is:

- Lukewarm response within middle management to teleconferencing 1) as an alternative to travel because travel is an integral ingredient of status and 2) because of the perception that teleconferencing is a new management toy.

Apart from the psychological problems, "technological" problems might stall the development of teleconferencing and be used as a smoke-screen to divert attention from invested motivations. These problems may include:

- The difficulties of security and encryption;

- The costs of establishing fixed teleconferencing facilities in various locations;

- The logistical nightmare of integrating remote pick-ups, local loops and in-house hookups within short turnaround time frames.

There will be resistance. However, if teleconferencing technology has any inherent merit, time will overcome this. My own view is that the cost of the technology will be the major problem, rather than resistance from middle management. When the technology is sufficiently developed so that virtually anyone can access and use a teleconferencing system at reasonable cost, organizational resistance will begin to ebb. It is a matter of *when* this will happen, not if.

NEW ORGANIZATIONAL ALLIANCES

These economic, technological and communications patterns ultimately converge around an "organizational" beacon. It is my view that all organizational problems stem from functional structure regardless of the content of the functions involved. As Harold Innis, Marshall McLuhan and others have indicated, when a new technology pokes its head into a culture, it ultimately has an impact on the structure and organization of that culture. Teleconferencing will be no exception. The question for organizational managers is: Who will manage teleconferencing? Who should be in charge of developing the use of the medium? Where in the organization does the teleconferencing expertise belong?

These are thorny questions because the answers may be different for each organization. Organizations should find expertise in telecommunications departments. However, does this mean that the telecommunications department should run and manage the teleconferencing system? Consider this: a teleconference involves two or more locations connected either by land lines, microwave or satellite (or a combination). The connection among locations is a telecommunications function. But what about the teleconference rooms at either end of the system (particularly if video conferencing is involved)? There are considerations of video cameras, lighting, multiple audio setups, room acoustics, furniture, etc. In many organizations, the in-house video production department would have expertise in this area. And what about the possible transmission of textual material from one location to another? This type of communication might involve word processing or data processing (i.e., office technology) personnel. Finally, there is the question of scheduling and administration. Should the telecommunications department schedule the facility? Or should this function be handled by a department more involved in the use of teleconferencing, rather than its technology?

A possible solution is that the adoption, implementation and development of teleconferencing should be launched as a *partnership* among all potential users and implementers, assisted when necessary by outside resources. This partnership is inevitable. Without new organizational alliances, teleconferencing will not find full expression.

There is the prospect that teleconferencing's growth will disturb the prevailing structure. Just as the telephone, photocopying machine and video communications do not belong to just one department, neither does teleconferencing. The technology is potentially global in impact; the management of it must follow similarly. Teleconferencing will work to its fullest when it is used for voice, data *and* video communications. There is tremendous opportunity here for video communications, telecommunications, data processing and various user departments to join together to make it work.

USING TELECONFERENCING: A MANAGER'S CHECKLIST

These macro-issues and problems aside, the development of a teleconferencing system must take into account many aspects. As the following checklist indicates, there are many problems to deal with.*

1) Location of equipment
    Proximity encourages usage
    Home turf encourages usage
    Locating equipment where people normally travel encourages usage
    Travel barriers discourage use

2) Need
    Do potential users recognize the need or is it recognized by outside evaluators and/or the boss?
    Do users want to meet this need?
    What's in it for the users?

3) Repair Service
    Is on-site technical repair or assistance available?
    Is technical support competent?

---

*This checklist is based heavily on the work of Martin Elton, John Carey, several outside suppliers and the author's experience.

4) Training of users
   Who will conduct it?
   How will it be conducted?
   Where should it be conducted?

5) Comfort
   What do users want in a teleconferencing situation?
   How will people interact in a teleconferencing environment?

6) Room design
   Room size? How many people will it accommodate?
   What kind of audio system?
   What kinds of cameras (if video conferencing)?
   TV monitor size?
   Acoustics, lighting, decor?
   User control systems?
   Table configuration?
   Peripherals: CRT's, facsimile, slow-scan?
   Electronic blackboard?
   Telephones?
   Transcription services?
   Reception area?

7) Implementation
   How long will it take?
   What are the hidden obstacles?
   What stages should the implementation process go through?
   How long will it take to debug the system?

8) Turnover in the user group
   How will this affect system use?
   How can user turnover be handled?

9) Security
   What is the need?
   Who will administer system security?
   What equipment is available to ensure information
      transmission security and privacy?

10) Early user involvement
   How can potential users become involved in the planning
      and implementation process?

Will this overcome users' potential inhibitions about the technology?

11) Scheduling
   Time zone differences?
   Who handles the scheduling?
   How much lead time is required?
   Preparing the users for the teleconference?
   Reserving time?

12) Uses of the teleconference system
   Management communications?
   Product promotion?
   Technical information exchange?
   Training?
   Public affairs?
   Professional development?
   Employee communications?
   Marketing?

Regarding the problems of managing the actual teleconference itself, the following items should be considered:

1) What kind of teleconferencing system is needed? audio, enhanced audio, partial-motion video, full-motion video?
2) What resources are available?
3) How much will it cost? Is it cost-effective?
4) Who will handle setting up the teleconference?
5) Setting up the time for the teleconference
6) Notifying participants
7) Identifying the subject of the meeting
8) Identifying the objective of the meeting
9) Developing an agenda; designing the structure of the agenda
10) Preparing the participants; training participants to use the camera and audio controls
11) Special considerations with respect to participant interaction (e.g., speaking too loudly or too softly, wearing loud clothes in the case of a video conference)

In all, the best antidote for drowning in a sea of teleconferencing problems is to practice preventive medicine: ask questions first, then take action.

TRANSCENDING THE PAST

The euphoria of equipment suppliers, consultants and futurists notwithstanding, there is an understated debate as to whether present and future teleconferencing systems will cause revolutionary changes in the use of existing communications systems.

As I indicated in *Managing the Corporate Media Center* (Knowledge Industry Publications, Inc., 1981), whenever a new communications medium comes along, there seems to be a tendency for people to take an "all or nothing" posture. But we did not stop speaking as soon as we learned how to write, or stop reading when television came along.

New technologies are additive, especially if they are able to absorb the characteristics and capabilities of the older technologies. For example, written symbols (such as the alphabet) are graphic representations of speech; the content of printing is written symbols; the content of photography is graphic renderings (painting, for example); the content of film is still photography moving at 24 frames per second; the content of television is film, still photography, print and speech; a video disc can likewise absorb all the communications characteristics of speech, print, photography, film and television. Finally, a teleconference can use all of the above media. In a sense, newer technologies, especially if they are more efficient and effective, seem to reflect a "bigger fish" metaphor. Video conferencing, especially, seems to be the latest biggest fish.

If teleconferencing is the contemporary bigger fish, does this mean that television, film, print or speech will wither away? History tells us no. Similarly, will teleconferencing in its various forms do away with travel and reduce energy costs? The answer at the moment appears to be that there is the potential for *some* portion of business travel to be curtailed, although as some authors suggest, there is the possibility that teleconferencing could stimulate more travel.

All economic justification aside, perhaps the real reason teleconferencing has a chance to thrive is that it transcends what geographical dispersion has created: too much distance between people working in the same organization. Teleconferencing seems to create a more efficient and, therefore, more effective meeting; it overcomes the space barrier and, therefore, the people barrier.

SOURCES OF INFORMATION

New technologies should be approached by using as many information sources as possible. The teleconferencing field is sufficiently developed so that many information sources are available.

A prime source of information is the supplier. Many satellite suppliers, turnkey video conference suppliers, producers, consultants, systems designers and other relevant suppliers are listed in the Appendix to this book.

Another information source is trade publications, including: *Satellite Communications* (Cardiff Publishing Corp. , Denver, CO), *Satellite News* (Phillips Publishing Inc., Bethesda, MD) and Eliot Gold's *TeleSpan Newsletter* (Altadena, CA). The Bibliography at the end of this book suggests additional magazines and newsletters. Seminars and conferences may be sponsored by such publications, and also by trade shows such as Video Expo, run by Knowledge Industry Publications, Inc.

Some of the books worth reading on telecommunications in general and teleconferencing in particular include: *The Wired Society,* by James Martin (Prentice-Hall, 1978); *Future Developments in Telecommunications,* also by James Martin (Prentice-Hall, 1977); *Telecommunications in the United States: Trends and Policies,* edited by Leonard Lewin (Artech House, 1981); and *Electronic Meetings: Technical Alternatives and Social Choices,* written by Robert Johansen, *et al.,* of the Institute for the Future (Addison-Wesley, 1979).

Three associations have sprung up: the International Association of Satellite Users (IASU), the Public Service Satellite Consortium (PSSC) and the International Teleconferencing Association (ITA).

The IASU, formed in 1980, is an independent nonprofit international trade association of suppliers and users supported solely by membership dues and member services. It provides conferences and training seminars, planning information exchange among members, periodic bulletins on members services and products, and even periodic bulletins on legislative and regulatory actions that significantly affect user interests. IASU members include Aetna Life and Casualty, AMAF Industries, Chase Manhattan Bank, Custom Cable Television, HI-NET Communications, National Oceanic and Atmospheric Administration, Oak Communications, Satellite Business Systems, Times-Mirror Cable Television, Todd Communications and Western Union Telegraph Co.

The PSSC is a nonprofit organization founded in 1975 that also packages video conferences. PSSC owns its own satellite uplink facility in Denver, CO, and a transportable uplink and downlink. As of spring 1981 the PSSC had handled over 100 video conferences and served such clients as the U.S. Army, the American Bar Association, the American Hospital Association and the United Methodist Church.

The ITA was formed in mid-1982, and is a professional organization for users, vendors and researchers working in the field of teleconferencing. The ITA plans to provide an exchange of information between teleconferencing users and vendors, and to act as a forum for the development of new applications and techniques for teleconferencing. Current members include Citibank N.A., Aetna Life and Casualty, IT&T, NEC America Inc., Colorado Video and The Darome Connection.

FOOTNOTES

1. Martin C.J. Elton, "The Practice of Teleconferencing," in *Telecommunications in the United States,* ed. Leonard Lewin (Dedham, MA: Artech House, 1981), p. 261.

2. Ibid., p. 259.

3. Kim E. Degnan, Jack T. Pottle and Paul I. Bortz, *Teleconferencing: Industry Overview; Economic Parameters; Public Broadcasting Potential* (Denver, CO: Browne, Bortz, & Coddington, 1981), p. viii.

4. *Applications Brief to the FCC* (Washington, DC: United States Satellite Systems, Inc., 1981), p. 32.

5. Alan B. Kamman and Diane Sargent-Pollock, "Telecommunications in the Nineties," *Telecommunications in the United States,* ed. Leonard Lewin (Dedham, MA: Artech House, 1981), p. 1.

ACKNOWLEDGEMENTS

I am deeply indebted to the following individuals for their contributions to this chapter: Martin C.J. Elton, New York University; John Carey, Greystone Communications; Susan Pereyra, Connex International; Greg Paulsen, VideoNet; Dick Jackson, Aetna Life and Casualty; Paul Pallino, Union Carbide Corp.; Jeff Dunne, AT&T; Herman Cotler, United States Satellite Systems; and Anne Haughton, RCA Americom.

4

The Status of Teleconferencing

by Ellen A. Lazer

A great deal is being said about teleconferencing, but there are few hard facts about who teleconferences, which teleconferencing modes are most popular and why, and whether users are satisfied or not with teleconferencing. The tremendous amount of media attention devoted to teleconferencing in the last several years consists primarily of single case studies or system descriptions. Most of the studies that have been conducted are proprietary to one institution and/or focus on specifics, such as the size of the market for one kind of teleconferencing service, which teleconferencing format works best for a certain application, or user perceptions of one kind of teleconference used for one kind of meeting.*

The teleconferencing survey conducted by Knowledge Industry Publications, Inc. and reported here is exploratory and broader in scope, since it focuses on the general needs and perceptions of actual teleconferencing users. Its aim is to track the general status of audio, video and computer conferencing, including:

- What experience business, government and nonprofit organizations are having with teleconferencing;

*Among the organizations active in research on teleconferencing are AT&T, the British Broadcasting Corp. (BBC), the Corporation for Public Broadcasting (CPB), the Defense Advanced Research Projects Agency (DARPA), Future Systems, Inc., the Institute for the Future, the International Association of Business Communicators (IABC), the National Aeronautics and Space Administration (NASA), the National Technical Information Services (NTIS), the Public Service Satellite Consortium (PSSC), Satellite Business Systems, Inc. (SBS) and Western Union.

- What purposes these teleconferences serve;

- Whether video conferencing will become commonplace;

- How teleconferencing is managed; and

- What benefits and problems are currently associated with teleconferencing.

Since the number of organizations involved with teleconferencing is growing rapidly, this survey is not intended as a comprehensive study of all
users. Rather, it is intended to augment the many specialized studies,
journalistic surveys and personal experiences that have contributed to our
knowledge of teleconferencing thus far.

HOW THE SURVEY WAS CONDUCTED

In spring 1982 Knowledge Industry Publications, Inc., sent a detailed
questionnaire (reprinted as Appendix 4A at the end of this chapter) to
1000 "Teleconferencing Managers" of the *Fortune* 1000, and to 1000
known nonbroadcast video users. We chose these lists in the hope that they
would elicit respondents who were knowledgeable and active in teleconferencing. One hundred seventy-three usable responses were received from 96
for-profit and 77 nonprofit organizations.*

As Table 4.1 shows, most of the responding for-profit companies are
large, with annual sales of at least $500 million, and are dominated by
manufacturers. Most have thousands of employees. Educational institutions dominate the nonprofit respondents (see Table 4.2), most of whom
have fewer than 1000 employees.

The titles of our respondents varied tremendously, from corporate telecommunications manager to video producer and media coordinator, and
even to training director, public relations manager and office services
manager. In nonprofit organizations and smaller companies, the respondent tended to work for a video or audiovisual department; in the larger
companies, he or she could just as well be in charge of communications or
telecommunications (see Table 4.3).

The balance of this chapter discusses the results of the survey, suggests
reasons for many of the findings and raises some issues and projections for
the future.

---

*These numbers, of course, are too small to lead one to other than tentative conclusions,
and cross-analysis of the items was held to a minimum.

**Table 4.1   For-profit Respondents by Main Function,
Sales Revenue and Number of Employees**

	No. of Respondents	Percent of Total
Function		
Manufacturing	63	66%
Financial services	22	23
Communications	6	6
Other	5	5
Total	96	
Annual Sales Revenues		
Under $50 million	6	6%
$50-$500 million	31	32
$500 million-$1 billion	20	21
More than $1 billion	37	39
NA	2	2
Total	96	
Number of Employees		
Under 1,000	7	7%
1,000-5,000	37	39
5,000-15,000	23	24
More than 15,000	29	30
Total	96	

NA: Not applicable.

THE USE OF TELECONFERENCING

Expenditures on Travel

Since the potential for reduced travel is the most frequently promoted rationale for teleconferencing—especially for two-way video conferencing —we asked how much companies spend on travel. As Table 4.4 shows, 56% of the for-profit companies say they spend more than $500,000 annually on business-related travel; nearly all spend more than $50,000. Not surprisingly, the nonprofit organizations report considerably lower travel expenditures: 65% spend less than $50,000 on travel annually, and only 26% spend from $50,000 to $500,000.

Table 4.2   Nonprofit Respondents by Main Function
and Number of Employees

	No. of Respondents	Percent of Total
Function		
Colleges and universities	35	46%
Hospitals and medical centers	17	22
Government, military and law enforcement agencies	12	16
Libraries	8	10
Public schools	5	6
Total	77	
Number of Employees		
Under 1,000	51	66%
1,000-5,000	21	27
5,000-15,000	3	4
More than 15,000	0	NA
NA	2	3
Total	77	

NA: Not applicable.

Past Experience with Teleconferencing

There is no question about the emphasis on audio conferencing. More organizations have used audio conferencing than any other form of tele-conferencing—about as many as have used all the other forms of teleconferencing combined. Enhanced audio ranks second in popularity among the for-profit respondents, while computer conferencing ranks second among the nonprofit respondents, especially the colleges. Figure 4.1 illustrates the relative experiences of organizations with five forms of teleconferencing. It should be noted that if one-way video conferencing is counted together with enhanced audio conferencing, this category will overshadow computer conferencing.

Not surprisingly, only 13% of our respondents (22 organizations) have experienced a two-way video conference, and the reasons for this are brought out later in this chapter. What is surprising is that 14% of our sample (21 organizations) indicated that they have done no teleconferencing at all—despite the fact that this group is very "literate" in communications technology. More than four out of five respondents have in-house video production facilities—nearly all with studios, editing facilities and full-time staff. In addition, two thirds of the respondents have video tape networks, and three fourths of the for-profit companies have a data or telecommunications network.

Table 4.3 Respondents' Job Titles by Type of Organization

| | For-profit | | | |
| | | | | |
Department	Small (less than $50 million revenues)	Medium ($50-$500 million revenues)	Large (more than $1 billion revenues)	Total
Media (video, TV, A/V)	18	7	16	41
Telecommunications/ communications	3	4	15	22
Information/data processing	3	1	2	6
Corporate communications/public relations	3	1	0	4
Administrative/ office services	5	3	1	9
Personnel/training and development	2	3	0	5
Other	1	0	0	1
NA	4	1	3	8
Total	41	20	37	96

| | Nonprofit | | | |
Department	Educational	Medical	Other	Total
Media (video, TV, A/V)	23	7	15	45
Instructional media	5	0	1	6
Telecommunications/ communications	2	0	2	4
Personnel/training and development	1	4	0	5
Other	3	4	7	14
NA	1	2	0	3
Total	35	17	25	77

NA: Not applicable.

Perhaps the biggest surprise of all was that the level of teleconferencing activity—and the kinds of teleconferencing being done—do not seem to have any correlation to the amount of travel expenditures. The most common form of teleconferencing today—regardless of the size of the

Table 4.4   Current Spending on Business-related Travel

Amount	For-profit Respondents	Nonprofit Respondents	All Respondents
Less than $50,000	4	50	54
From $50-$500,000	33	20	53
More than $500,000	54	0	54
No answer	5	7	12
Total	96	77	173

organization's travel budget, and regardless of whether it is a for-profit or nonprofit organization—remains audio conferencing.

Video Conferencing

Since video conferencing is the newest and most expensive form of teleconferencing, we asked whether companies who have not used it in the past plan to do so in the future—and indeed they do. Eighteen respondents plan to video conference in the following year, and 43 plan to do so over the next two years. Video conferencing is on the drawing boards, if not yet in the budgets, of many organizations.

Organizations who do not plan to video conference in the next two years were asked why. As Table 4.5 indicates, the most frequently given answer was cost, followed by "the benefits are not clear" and "no corporate needs have developed." It was interesting that almost no one checked the statements "video conferencing is not technologically sound" or "managers would be reluctant to give up travel."

Table 4.5   Reasons for Not Video Conferencing

Reason	For-profit	Nonprofit	Total
Too expensive	25	19	44
Benefits not clear	18	10	28
No needs have developed	16	12	28
Top management uninterested	11	8	19
Not as effective as face-to-face meeting	9	7	16
Unnecessary	1	5	6
Other[1]	6	4	10

[1]For example: equipment is used for other purposes, lack of knowledge, not technologically sound (two responses), reluctance to give up travel (two responses).

Note: More than one response was allowed.

Figure 4.1 Experience with Teleconferencing

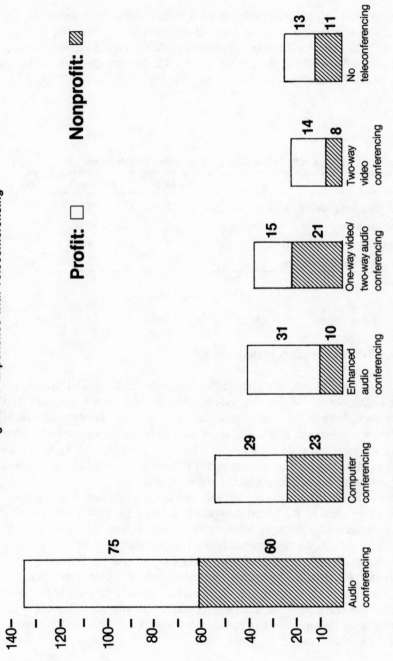

Profit: ☐   Nonprofit: ▨

Note: Some respondents checked more than one form of conferencing.

How Companies Video Conference

Nearly half of the respondents who video conference today do so through a commercial service such as Picturephone® Meeting Service (PMS), Public Service Satellite Consortium (PSSC), etc. (other services are listed in the Appendix at the end of this book). Twenty-nine percent have in-house teleconferencing facilities, and 16% rent facilities (see Table 4.6).

Table 4.6   How Companies Video Conference

	For-profit	Nonprofit	Total
Have in-house facilities	12	15	27
Rent facilities	5	10	15
Use outside consultant	2	4	6
Use commercial service	23	22	45

Future Teleconferencing Plans

As Table 4.7 shows, change is afoot regarding audio and video conferencing. One fourth of the respondents plan to do a demonstration video conference; nearly as many plan to expand their present telephone conferencing systems, and many show interest in setting up an audio conferencing system or doing *ad hoc* video conferencing—though just as many plan no video conferencing at all. Fifteen organizations say they will establish a formal teleconferencing department.

One indicator of a company's intentions is the equipment it already has or plans to obtain. We asked companies with an interest in an in-house video conferencing network whether they buy or rent certain equipment (see Table 4.8). Sixty-five respondents, for example, currently own video equipment, 20 plan to purchase it and 11 plan to rent it. Similarly, most respondents currently own audio equipment, but more plan to rent it in the future rather than purchase it. Only seven organizations currently own receive-only earth stations, and only 23 plan to purchase or lease one—yet 49 (primarily the nonprofit organizations) would buy or lease a transmit-receive earth station.

Table 4.7   Future Audio and Video Conferencing Activities

	For-profit	Nonprofit	Total
Plan demonstration video conference	25	18	43
Expand telephone conference system	31	8	39
Set up in-house video conference network	26	8	34
Use outside service to:			
video conference regularly	4	1	5
do an *ad hoc* video conference	18	11	29
do a demonstration	15	4	19
Set up audio conference system	26	6	32
No video conferencing	12	16	28
Establish teleconferencing department	10	5	15
No audio conferencing	1	7	8
Other[1]	2	3	5
No answer	10	20	30

[1]For example: continue occasional teleconferencing; conduct pilot audio conference; expand freeze-frame teleconferencing.

Teleconferencing Applications

The most popular applications of audio and video conferencing among for-profit companies are employee communications, followed by professional/technical consultation and sales meetings. Education is the number one application among nonprofit organizations; second is consultation and third is employee communications. In addition to these categories, which were proposed in the questionnaire and are shown in Table 4.9, respondents indicated use of teleconferencing for such applications as board meetings, financial and project reviews, problem resolution, alumni meetings, scheduling and staff meetings, security, technical briefings and even job interviews.

HOW TELECONFERENCING IS MANAGED

Who's in Charge?

As is common when a new function emerges, it is not certain where teleconferencing fits in an organization's structure. We asked, "Which department in your company is charged with the responsibility of managing tele-

Table 4.8   Equipment Purchase Plans for In-house Video Conferencing

Equipment	For-profit	Nonprofit	Total
Video Equipment			
Currently own	27	38	65
Currently rent	8	0	8
Will purchase	13	7	20
Will rent	11	0	11
Audio Equipment			
Currently own	22	32	54
Currently rent	9	0	9
Will purchase	9	2	11
Will rent	8	0	8
Earth Stations			
Currently own TVRO	3	4	7
Will purchase TVRO	3	6	9
Will lease TVRO	10	4	14
Will buy or lease transmit/ receive earth station	19	30	49
No answer	42	30	72

conferencing," and the responses indicate that it is primarily concentrated in communications, media/AV and video/TV departments (see Table 4.10). However, a significant number of companies assign teleconferencing to departments such as training, office services, data processing and marketing. This is probably because a department like training or marketing sees its own need to hold a teleconference—which means that several departments in one organization could be running their own teleconferences quite independently of each other. The data do indicate that teleconferencing is being centralized, and that communications seems a logical home. Whether teleconferencing will emerge as an independent department, or one attached to a video, communications or other function, remains to be seen.

Budgets

Only six respondents indicated that their organizations have a separate budget for teleconferencing, and only 27 respondents said that their organizations identify teleconferencing expenditures in the department that runs it: e.g., general administration, telecommunications, video production, marketing, employee communications, training, human resources, corporate relations, investor relations, public affairs. We may speculate

Table 4.9   Audio and Video Conferencing Applications

Application	For-profit	Nonprofit	Total
Employee communications	50	17	67
Audio	36	13	49
Video	14	4	18
Education, seminars	22	38	60
Audio	12	23	35
Video	10	15	25
Professional consultation	33	22	55
Audio	24	14	38
Video	11	2	13
Sales meetings	33	1	34
Audio	23	1	24
Video	10	0	10
External communications	23	10	33
Audio	19	7	26
Video	4	3	7
Dealer/customer information	20	4	24
Audio	15	2	17
Video	5	2	7
Sales training	19	2	21
Audio	9	2	11
Video	10	0	10
Dealer/customer marketing	21	0	21
Audio	16	0	16
Video	5	0	5
Stockholder meetings	4	2	6
Audio	2	0	2
Video	2	2	4
Other*	9	8	17
Audio	5	3	8
Video	4	5	9
All audio	161	65	226
All video	75	33	108

*These include public service announcements, board meetings, acting as host sites for outside teleconference users, job interviews, staff reports, security, etc.

that this is because the less expensive audio conferencing falls under general overhead, at least at the present time, while the more expensive video conferencing is likely to be charged back to a particular department. Once a video conference network is established, of course, it might become less of an extraordinary expense and may take its place with the audio budget.

Table 4.10   Department Supervising Teleconferencing

	For-profit	Nonprofit	Total
Communications	43	8	51
Media/AV	12	35	47
Video/TV	27	17	44
Training	14	6	20
Administration/office services	14	2	16
Data processing/MIS	14	0	14
Marketing	11	2	13
PR/corporate communications	8	1	9
Conferences and meetings	4	4	8
Travel/transportation	2	0	2
Other'	8	6	14
NA	11	16	27

'Educational services, telecommunications, engineering/R&D, employee relations, security, continuing education.
Note: More than one response was allowed.

As Table 4.11 shows, the responding nonprofit and for-profit organizations are at opposite poles when it comes to all money questions, even pie-in-the-sky ones; the average "ideal" budget for teleconferencing among nonprofit organizations is only $48,000, compared to $383,000 for the for-profit companies. In 1981, the nonprofit organizations spent an average of $5500 on teleconferencing; the for-profit companies spent an average of $144,000. All the respondents expect teleconferencing expenditures to grow; by 1985 the for-profit companies expect to spend more

Table 4.11   Median Estimated Expenditures on Teleconferencing

Year	For-profit	Nonprofit
	Ideal budget	
	$383,000	$48,000
	Actual Proposed Budget	
1981 budget	144,000	5,500
1983 estimate	453,000	10,000
1985 estimate	728,000	21,000
1990 estimate	1,406,000	41,500

Note: Based on 35 responses.

than $700,000 on teleconferencing, and they expect to spend well over a million dollars by 1990.

If the respondents were given $100,000 to spend on telecommunications equipment, 40% of the for-profit companies would spend it on setting up a permanent video conferencing network and another 30% would spend it on modernizing the telephone conferencing system. Sixty-two percent of the nonprofit companies, on the other hand, would use the money to obtain additional video equipment; another 32% would buy a television receive-only earth station (TVRO).

We asked respondents to rank 22 suggested expenses associated with the budget for an *ad hoc* single event video conference. Respondents were asked to indicate whether the cost of each item was considered high, medium or low, and whether the activity would be done in-house or with an outside vendor. Among the findings, shown in detail in Table 4.12, are:

- Most frequently ranked as low cost are participants' expenses, telephone calls, sound systems and catering.

- Most frequently ranked as high cost are transmission and TV reception at the sites, followed by participants' time and technical facilities.

- An outside vendor is more likely to be used for such services as transmission and specially installed telephones, and TV reception at the sites. A vendor is least likely to be called upon for planning and scripting, preparation of visuals, or for technical facilities (such as studio equipment) and meeting site management.

PERCEIVED ADVANTAGES AND DISADVANTAGES OF TELECONFERENCING

The most frequently cited advantages of teleconferencing are to save travel time and to make meetings more cost-effective—these two reasons were listed by 80% of our respondents. Other important advantages of teleconferencing listed in Table 4.13 are to enable more people to attend more meetings, and to communicate important information quickly. Several respondents told us that they teleconference in order to improve productivity, to provide a broader base of management control and to generate quicker decisions. One respondent wrote simply, "There are no disadvantages to teleconferencing."

As Table 4.13 shows, the major disadvantage of teleconferencing is cost. Video conferencing, in particular, is just too expensive. It's important to

Table 4.12   Breakdown of Teleconference Costs

| | Cost | | | How done | |
Item	High	Medium	Low	In-house	Outside Vendor
Participants' time	33	47	26	NA	NA
Participants' expenses	15	49	46	NA	NA
Planning & scripting	22	46	33	87	17
Outside talent	25	29	36	34	48
Visual preparation	18	69	31	85	17
Production time	28	54	30	84	17
Production expenses	20	54	37	78	23
Technical facilities	32	48	30	73	28
Technical personnel	27	61	23	69	34
Transmission	52	39	10	14	77
Special telephones	16	34	46	26	68
Interactive Q/A	13	36	30	23	48
Telephones	9	42	49	52	39
Network organization	18	43	31	53	39
Meeting site surveys	13	39	38	55	39
Meeting site rental	26	41	35	42	52
TV reception	40	43	19	22	73
Image display	23	48	29	38	58
Sound system	12	42	43	53	41
Catering	6	40	45	37	48
Meeting site-technical	9	53	31	54	33
Meeting site-logistics	9	17	34	56	29

NA: Not applicable.

note than nearly an equal number of respondents think that teleconferencing is less effective than face-to-face meetings, and rate this, too, as a major disadvantage. Other disadvantages suggested include: old habits change slowly; there is too much show business associated with performances on a video conference; it's hard to watch a video screen for a long time; people hesitate to interact on a telecommunications medium; the benefits aren't clear; management is not interested. We were surprised that few find the technology not reliable, or hesitate to give up travel.

CONCLUSION

This exploratory survey suggests that teleconferencing is in place at most large organizations. Audio conferencing is particularly widespread. Augmented audio, computer and one-way video conferencing are common enough that growth in these modes seems inevitable. Two-way video

Table 4.13   Advantages and Disadvantages of Teleconferencing

	Rank	For-profit	Nonprofit	Total
Advantages				
Save travel time	1	36	25	61
	2	32	18	50
	3	13	19	32
Cost-effective	1	36	22	58
	2	37	16	53
	3	24	20	44
Can communicate quickly	1	24	12	36
	2	9	7	16
	3	15	5	20
More people can meet	1	11	24	35
	2	20	19	39
	3	47	14	61
Unify organization	1	7	8	15
	2	7	8	15
	3	9	4	13
Improve communication	1	5	1	6
	2	14	7	21
	3	7	6	13
Disadvantages				
Too expensive	1	39	29	68
	2	17	17	34
	3	15	10	25
Less effective	1	21	18	39
	2	30	26	56
	3	17	9	26
Loss of executive travel	1	10	9	19
	2	20	14	34
	3	37	25	62
Too much trouble	1	5	2	7
	2	12	12	24
	3	15	10	25
It's a gimmick	1	4	6	10
	2	6	3	9
	3	15	8	23

conferencing, however, faces serious obstacles, primarily those of cost and user perceptions of effectiveness.

Information exchange from one source to many points—whether among employees, colleagues or students—is the common denominator of the particular applications most widely cited by our survey. New uses are rapidly emerging, however, particularly for video conferencing (see Chapter 8).

Teleconferencing does not yet have a "home" in most organizations; we saw that it may be within the purview of a communications or video function, or a host of others; nor does it have single budget line in most organizations.

Teleconferencing is perhaps beginning to be appreciated for its effectiveness as one of many communications media, rather than as a substitute for a face-to-face meeting or as "merely" a means of saving time and money. The primary problems are basic and deal with cost and effectiveness; the human issues of becoming comfortable with technology are not yet paramount.

The number of suppliers is mushrooming and the number of organizations who audio or video conference frequently is growing. It seems certain that teleconferencing increases the amount, and possibly the quality, of many forms of business and informational communications—and decreases the time spent physically away from the workplace. As more institutions pay special attention to individual productivity and gain more experience with the various teleconferencing technologies, and as the costs of teleconferencing decrease as the technology becomes more accessible, the scenario shown in this chapter may change considerably.

Appendix 4A: Teleconferencing Questionnaire

Your job title: _____

Your name (optional): _____

Your company or organization (optional): _____

Address (optional): _____

1. Please characterize the size of your company:
   a) _____ under $20 million in sales revenue
      _____ from $20 million - $50 million in sales revenue
      _____ from $50 million - $500 million in sales revenue
      _____ from $500 million - $1 billion in sales revenue
      _____ more than $1 billion in sales revenue

   b) _____ under 1000 employees
      _____ from 1000-5000 employees
      _____ from 5000-15,000 employees
      _____ more than 15,000 employees

2. Your company is primarily engaged in the following (check all that apply):
   _____ manufacturing (please specify: _____)
   _____ services (please specify: _____)
   _____ education (please specify: _____)
   _____ nonprofit (please specify: _____)
   _____ communications (please specify: _____)
   _____ health care (please specify: _____)
   _____ other (please specify: _____)

3. Your company currently spends approximately the following on business-related travel each year (travel, lodging and entertainment expenses for meetings, training, seminars, etc.):

   _____ less than $50,000
   _____ from $50,000-$500,000
   _____ more than $500,000

4. Has your company used any of these forms of teleconferencing in the past?

   _____ computer to computer conferencing
   _____ voice (telephone) conferencing
   _____ slow scan video with audio: one-way _____ two-way _____
   _____ two-way video conferencing
   _____ one-way video, two-way audio
   _____ audio conferencing with video
   _____ other (please specify: _____)

5. For what purposes has your company used audio conferencing? Video conferencing? Please check all that apply:

   a. _____ sales training: _____ audio _____ video
   b. _____ sales meetings: _____ audio _____ video
   c. _____ dealer/customer communications (informational):
          _____ audio _____ video
   d. _____ dealer/customer sales (marketing): _____ audio _____ video
   e. _____ employee communications: _____ audio _____ video
   f. _____ stockholders' meetings: _____ audio _____ video
   g. _____ external communications: _____ audio _____ video
   h. _____ professional/technical/medical consultation (please explain:
          _____): _____ audio _____ video
   i. _____ educational, seminars, etc. (Please explain: _____)
          _____ audio _____ video
   j. _____ other (please specify: _____)
          _____ audio _____ video

6. If your company has not used video teleconferencing in the past, does it plan to do so in the: _____ next year _____ next two years.

7. If your company does not plan to video conference within the next two years, why not? (You may check more than one)

   _____ too expensive
   _____ benefits not clear
   _____ no corporate needs have developed
   _____ unnecessary
   _____ reluctance to give up travel
   _____ not as effective as face-to-face meetings
   _____ top management uninterested in it
   _____ not technologically sound
   _____ other (please specify: _____)

8. In the next two years, does your company plan to:

   _____ establish a formal teleconference department
   _____ do a demonstration video teleconference
   _____ set up an in-house video teleconferencing network
   _____ contract with outside service to provide video teleconferencing:
          _____ on a regular basis
          _____ as a demonstration (pilot) project
          _____ for an "ad hoc" special purpose teleconference
   _____ plan no video teleconferencing activity whatsoever
   _____ set up a telephone conferencing system
   _____ expand your present telephone conferencing system

_____ plan no telephone conferencing whatsoever
_____ other (please specify: _____)

9. Do you presently have an in-house video production facility?
_____ Yes _____ No
(If yes, indicate: _____ studio _____ editing facilities _____ full time staff)

10. Do you presently "network" programming by video tape distribution?
_____ Yes _____ No
If yes, to how many points _____
What format? _____

11. Do you currently have an in-house:
Data network _____ Yes _____ No
    Size _____
Telecommunications network _____ Yes _____ No
    Size _____

12. If you have or plan to set up an in-house video teleconferencing network, how have/will you set it up?

a) _____ currently own video equipment _____ will purchase
   _____ currently rent video equipment _____ will rent
b) _____ own audio equipment _____ will purchase
   _____ rent audio equipment _____ will rent
c) _____ own "receive only" earth station _____ will purchase
   _____ will lease "receive-only" earth station
   _____ will buy or lease transmit/receive earth station

13. Which department in your company is charged with the responsibility of managing teleconferences? (Check all that apply)

_____ video service   _____ marketing/sales
_____ media department   _____ public relations
_____ data processing/MIS   _____ travel department
_____ office services   _____ conferences & meetings
_____ communications   _____ training
_____ other (please specify: _____)

14. How does your company presently manage its video conferencing? (Check as many as apply)

_____ have installed in-house teleconferencing facilities
_____ rent facilities
_____ an outside consultant sets it up
_____ use a commercial service (please specify: _____)

_____ Picturephone Meeting Service
_____ PBS
_____ PSSC
_____ HiNet
_____ Professional teleconferencing company (please specify:
_____)

15. Have you been approached by _____ have you approached _____ vendors interested in setting up a teleconferencing system for you?

_____ Yes _____ "ad hoc" (one-time/pilot) _____ permanent _____ No
type of vendor _____

16. Have you been approached by commercial telecommunications services, such as Picturephone, SBS?: _____ Yes (please specify: _____)
_____ No

17. Does your company currently have a separate budget for teleconferencing?
_____ Yes (please specify: _____) _____ No

18. If no, are teleconferencing expenditures identified in the department(s) doing it? _____ Yes   In which department(s) budget? _____
_____ No

19. Your ideal annual budget for a teleconferencing department in your company would be: $ _____

What was your teleconferencing expenditure in 1981? $ _____

Realistically, your company will probably spend how much in 1982 on teleconferencing? $ _____ How much in 1983? $ _____
1985? $_____ 1990? $ _____

20. If you have $100,000 earmarked for telecommunications equipment, how would you rank the following expenditures:

_____ I'd add video equipment to our existing facilities
        (please specify: _____)
_____ I'd buy a TVRO (Television Receive-only Earth Station)
_____ I'd begin to set up a permanent video teleconferencing network among
        our sites
_____ I'd install communicating word processors
_____ I'd modernize our telephone conferencing system
_____ Other (please specify: _____)

21. When preparing the budget of an "ad hoc" video teleconference, which of the

following do/would you include? Please indicate estimated cost — H (high), M (medium), L (low) and whether in-house (I.H.) or professional vendor (P.V.).

	Estimated Cost			Source	
	H	M	L	I.H.	P.V.
Participants' time	___	___	___	___	___
Participants' expenses	___	___	___	___	___
Writing—content, planning	___	___	___	___	___
Outside talent (host, announcer, actors)	___	___	___	___	___
Visual preparation (slides, tapes, films)	___	___	___	___	___
Production team's time	___	___	___	___	___
Production team's expenses	___	___	___	___	___
Technical facilities (studio, cameras)	___	___	___	___	___
Technical personnel	___	___	___	___	___
Transmission costs (AT&T, satellite time)	___	___	___	___	___
Specially installed telephones	___	___	___	___	___
Interactive Q/A systems (2-way radio)	___	___	___	___	___
Cost of phone calls	___	___	___	___	___
Network organization	___	___	___	___	___
Meeting site surveys	___	___	___	___	___
Meeting site rental	___	___	___	___	___
TV reception at sites (earth stations, telco)	___	___	___	___	___
Image display at sites (TV receivers/ projectors)	___	___	___	___	___
Sound systems	___	___	___	___	___
Catering expenses	___	___	___	___	___
Meeting site management-technical	___	___	___	___	___
Meeting site management-logistic	___	___	___	___	___
Other_____	___	___	___	___	___

22. What do you see as the major advantages that teleconferencing offers your organization? Please rank the following from 1-8, with 1 being most advantageous.

_____ save travel time
_____ enable more people to attend more meetings
_____ make meetings more cost-effective
_____ give the media department a chance to contribute more effectively
_____ generally improve communications
_____ communicate important information very quickly
_____ help unify separate divisions, departments, regions
_____ other (please specify: _____)

23. What do you see as the major disadvantages? Please rank as above, with 1 being the most disadvantageous.

_____ too expensive
_____ less effective than person to person meetings
_____ loss of travel opportunities for executives
_____ loss of travel opportunities for staff
_____ fear of security leaks
_____ too much trouble to investigate and set up
_____ too gimmicky, not reliable
_____ other (please specify: _____)

24. What sources do you/would you use to learn more about teleconferencing?

_____ Trade magazines/journals (please specify: _____)
_____ Books, directories (please specify: _____)
_____ Vendors (please specify: _____)
_____ Consultants (please specify: _____)
_____ Trade shows, conventions, professional meetings (please specify:

_____)
_____ Colleagues in other organizations
_____ Seminars (please specify: _____)
_____ Other (please specify: _____)

25. Would you want specific educational programs on the subject of teleconferencing? _____ Yes _____ No. If yes, would you prefer these to be _____ in house or _____ professional seminars. Should they be targeted to _____ staff personnel _____ middle management _____ top management.

5

The Economics of Teleconferencing

by Bonnie Siverd

Few business decisions are made without at least a cursory nod to the bottom line. In the case of teleconferencing, scrutiny of the costs is all the more intense. The technology is new and often unfamiliar; the budgets involved can be substantial. Indeed, whether a teleconference takes place at all, or what kind is chosen, is frequently a function of the total bill.

IDENTIFYING THE VARIABLES

What does it cost to organize a teleconference? How much time and money can be saved by doing so? While these questions are of paramount importance to users, there are no clear-cut answers.

Unlike a simple memo or phone call, teleconferences are subject to an immense spread in nearly every element of their cost. The meeting itself, for example, could range from an off-the-cuff discussion of an engineering issue during an audio-only session to a complex video production featuring professional entertainment. The chief variable is the kind of conference selected—audio, one-way video or two-way video. Other cost factors include the number of locations, time of day and duration of the meeting, size of the audience, number of participants and degree of sophistication desired. And they extend to details as minute as the size of the signs, if any, that embellish the set, if any, in a video meeting.

What is more, the same budget can yield two vastly different teleconferences, depending on the quality of the production involved. James W. Johnson, president of TeleConcepts in Communications Inc. (New York, NY), likens the situation to that of designing a home. For a given price, a

Users can keep costs down by choosing a simple set. The one shown above can often be obtained for less than $1000. Courtesy VideoNet.

builder can put up a five-room bungalow or a 10-room house, each with vastly different attributes.

The simplest—and least expensive—teleconference is a prearranged telephone call between two parties. It can cost as little as the charge for a local call. For a more elaborate telephone meeting, says Susan Pereyra, president of Connex International, Inc. (Danbury, CT), a good rule of thumb is to budget $48 an hour per location.

Adding video will send expenses soaring. While video conferencing costs vary considerably, Adele Brown, director of marketing for the JMP Videoconference Group of Jack Morton Productions Inc. (New York, NY), notes that a typical two- to three-hour conference can usually be held for $100 or less per person. Because the entire production and uplinking process must be duplicated for each originating site, the cost of two-way video conferences is appreciably higher.

Within each of these categories, costs are frequently influenced by a meeting's locations. The direct dial charges between sites are a significant item in an audio budget. In video conferencing, says David Badoud, eastern regional manager for VideoNet (Woodland Hills, CA), equipment costs vary regionally. The proliferation of teleconferencing equipment in the Northeast industrial areas generally has kept rates competitive. In Texas and the South, by contrast, the relative scarcity of gear has led to higher fees. In addition, Badoud points out, cities on the edge of the footprint,* such as Seattle, Boston and Miami, require more sophisticated or larger equipment to maintain signal quality. As a result, transmission to these areas often entails higher costs.

Even within cities, the cost of services can differ markedly. To set up an earth station at the Century Plaza Hotel in Los Angeles, Badoud explains, cost between $1000 and $1300 in spring 1982. But terrestrial interference from surrounding buildings would boost the bill for a similar operation at the downtown Bonaventure to from $2400 to $4000 because a portable satellite receiver is all but impossible to use.

BASIC COST COMPARISON

Within the broad range of individual teleconferencing costs, certain categories of expenses recur. In determining the budget for a particular meeting, the following are the minimum elements to be considered:

- *Preproduction.* Costs of developing and publicizing the program, providing visual support materials and the like.

- *Production.* Fees for producing the session, including studio time.

- *Transmission.* Fees for sending the session, including speakerphone, satellite time and connections.

- *Receiving sites.* Costs of receiving the program, the meeting space and related items.

No one budget, of course, can indicate what teleconferencing will cost, since every meeting is unique. But for illustrative purposes, Table 5.1 outlines the cost of a similar but not identical meeting held four different ways. In all cases, a three-hour session takes place at a corporation in Stamford, CT. Approximately 60 participants from each of 21 U.S. cities (20 receiving sites plus Stamford) attend, either in person, as in Example A, or through various teleconferences, as in Examples B, C and D.

---

*The geographical area at which a satellite signal is aimed is known as a satellite's "footprint."

Table 5.1   Comparison of Meeting Costs:
Three-hour Meeting, 21 Cities, 60 Participants/City

	Example A: Face-to-Face meeting	Example B: Audio conference	One-way Video/ Two-way Audio	
			Example C: Video conference (using HI-NET system)	Example D: Custom-built video conference
Preproduction				
Visual inserts, miscellaneous graphics[1]	$ 25,000	NA	$ 10,000	$ 10,000
Production				
Speakerphone-type device with four microphones (rental and transportation)	NA	$ 81	NA	NA
Lighting	NA	NA	3,000	3,000
Crew and mobile unit	NA	NA	10,000	10,000
Teleprompter[2]	NA	NA	750	750
Director/producer/speech coach	NA	NA	5,000	5,000
Moderate set[2]	NA	NA	2,000	2,000
Rehearsal	NA	NA	8,000	8,000
Return audio	NA	NA	1,000	1,000
Transmission				
Bridge[3] (3-hour meeting, 1-hour test)	NA	1,680	NA	NA
Calls to bridge[4] (3-hour meeting, 1-hour test)	NA	2,352	NA	NA
Uplink (3-hour meeting, 2-hour test)	NA	NA	250	250
Satellite time[5] (3-hour meeting, 2-hour test)	NA	NA	5,000	2,500
AT&T charges	NA	NA	4,400	4,400
Receiving sites				
Speakerphone-type devices (rental and transportation)	NA	1,000	NA	NA
Downlinks	NA	NA	23,000	28,000
Projection equipment and services	NA	NA	15,000	15,000
Facilities	NA	NA	NA	NA
Refreshments (coffee and lunch, $10 per head)	12,600	12,600	12,600	12,600

Table 5.1   Comparison of Meeting Costs:
Three-hour Meeting, 21 Cities, 60 Participants/City (cont.)

Travel costs				
Air travel [6]	61,000	NA	NA	NA
Lodging	102,000	NA	NA	NA
Local transportation[7]	63,600[7a]	NA	19,200[7b]	NA
Meals[8]	60,000	NA	NA	NA
Total Cost	$324,200	$17,713	$119,200	$102,500
Cost Per Participant	$257	$14	$95	$81

NA: Not applicable.

NOTES:

Costs are as of May 1982. Examples assume sponsoring organization writes scripts, handles any necessary publicity, mailings, and other organizational tasks and provides a host or hostess at each site. In all cases, specific rates would depend on the precise locations used and the date and time of the meeting.

The face-to-face meeting travel costs were based on the least expensive round-trip economy air fare to La Guardia Airport listed in the *Official Airline Guide;* round-trip travel between Stamford and the airport in a local limousine service ($33) plus an estimated $20 in local travel costs, and one night's stay at a Stamford hotel at the least expensive corporate group rate of $85 per night.

The audio conference costs were estimated by Connex.

The video conferencing costs were estimated by VideoNet, except for quote in Example C on HI-NET downlinks from Holiday Inn. In Example D, the downlink charge assumes site surveys as necessary, technicians at each location, installation of 19 portable earth stations plus one AT&T connection. The projection equipment in both Examples C and D includes two 60-inch diagonal projection screens as well as backup equipment.

[1]Example A: slides and overhead projectors; Examples C and D: electronically produced video tape inserts.

[2]Some companies may incur costs for these items at a face-to-face meeting.

[3]$20 per hour x 21 sites x 4 hours.

[4]$28 per hour (an average long distance call estimated by Connex) x 21 sites x 4 hours.

[5]Satellite time for HI-NET can vary from $600-$2500 per hour.

[6]Based on round-trip economy air fare.

[7a]Based on $33 round-trip limo to and from airport, cabs to local airport, $20 ($53 x 1200 = $63,600).

[7b]Based on average round trip to receiving site of 80 miles, at 20 cents per mile x 1200 people. Local Stamford employees not included.

[8]Based on $50 per person meal allowance x 1200 people = $60,000.

Design and construction of an elaborate set such as the one shown above could easily cost $15,000. Courtesy TeleConcepts in Communications, Inc.

The face-to-face meeting, Example A, is held in the corporate auditorium, both to hold down costs and to help people feel comfortable. A simple video presentation is used. In the two-way audio conference, Example B, the main group speaks from Stamford and two additional speakers join from other locations, but there are no video inserts.* With the one-way video/two-way audio meetings, Examples C and D, the production originates from Stamford and contains video inserts comparable to those in Example A. In Example C, the program is beamed to viewing sites in the Holiday Inn HI-NET system. In Example D, the program is transmitted to various corporate branch offices where special receiving equipment has been installed.

The differences in costs are striking. Air travel and lodging bills alone add $163,000 to the tab for the face-to-face meeting in Example A, bringing its bill to $324,200. That compares with just over $100,000 for the video conferences in Examples C and D. On a per-participant basis, the spread is equally marked. Under the circumstances described, it would cost less than $100 a head to hold a video conference, while expenses for each person attending the face-to-face session would come to $257.

Because the audio meeting in Example B was produced on a simpler scale, with no video elements, its cost was even lower. The fees totaled just $17,713, or $14 per person.

COST-EFFECTIVENESS OF TELECONFERENCING

For many teleconferencing users, whether business or nonprofit organizations, the main attraction of this new technology is its potential to reduce costs. Chief among these are the soaring bills for travel, including air fares, and hotel and food costs. U.S. industry alone spent $21 billion on domestic business travel in 1980, according to Paul C. Sheeline, chief executive officer of Intercontinental Hotels Corp. An additional $12 billion—representing more than 10 million business trips—was devoted to international business travel. Indeed, of the 173 organizations responding to the questionnaire described in Chapter 4, 53 reported business-related travel budgets of between $50,000 and $500,000 a year and 54 laid out more than $500,000 annually.

Teleconferencing frequently can reduce at least some of these costs. A typical two- to three-hour teleconference for between three and 10 persons, reports Connex, may save from $1500 to $4000 in travel costs (assuming travel costs of $400 to $500 per person). The more frequently teleconferen-

---

*Other possibilities include: mailing visuals in advance; using a remotely controlled slide or fiche projector; mailing slides in advance and calling them up by number. These will add to the costs—but may make the meeting much more effective.

cing displaces travel, the greater the benefits. In the third quarter of 1981, for example, Honeywell (Minneapolis, MN) calculates that use of its six audio conferencing rooms cut travel costs by a net average $25,100 a month, notes Rick Whiting, senior project administrator and former manager of teleconferencing projects. Video conferencing can likewise reduce travel expenses. When Allied Van Lines (Chicago, IL) replaced a series of regional meetings with a teleconference in 1981, reports Kathryn Bradford, manager of meetings and conventions, the travel expenses for corporate support staff for that single project fell from $54,000 to $13,000. (The Allied teleconference is described in greater detail below.)

Such out-of-pocket savings can be readily tracked. But many of the cost "savings" associated with teleconferencing are intangible. A prime example is the increased number of hours that can be devoted to work instead of traveling. Honeywell's Whiting figures that such gains in time save Honeywell another $25,000 or so a month in salaries and fringe benefits—or roughly as much as its out-of-pocket travel displacement savings.

Increasingly, however, businesses are focusing on the more indirect cost savings discussed in Chapter 3. It is impossible to quantify the benefits of faster decision making or increased efficiency that result from a less harried schedule.

DOMESTIC TELECONFERENCING: COST COMPARISONS

Audio Conferences

An audio or voice-only meeting is the most frequent form of teleconferencing. Because it does not entail costly video transmission, it is also the least expensive.

As in all types of conferencing, the price will reflect the elaborateness of the session. The final cost of an audio meeting will be influenced primarily by the number of participants per site, the number of locations, duration of the meeting and use, if any, of graphic, video or written support material.

The simplest two-party telephone call costs little more than the direct dial charges involved. A somewhat more complex meeting can be arranged through an AT&T operator (at an additional cost, depending on the number of people, locations and duration).

One such meeting was held by the American Council on Education (ACE) in summer 1981. Twelve members of its board of directors, each with a telephone, assembled in an ACE conference room in Washington, DC. Twenty-three other board members joined from 23 additional sites in the U.S. The bill for the 35-minute session, including installation charges for necessary phone lines to the ACE boardroom, came to $680. Costs per participant averaged $19.

Audio conferences can also be arranged through vendors such as Connex, Darome Inc., and Kellogg Corp. In addition to direct dial costs for a meeting, these organizations charged from $16 to $20 per hour per location for their bridging services in May 1982.

In preparing estimates for such sessions, Jonna Lee Masters, manager of educational planning and development with Datapoint Corp. (San Antonio, TX), uses what she calls "$20-$20-$20" as a rule of thumb. Preparation costs, she estimates, average $20 per site. This figure includes bills for printed handouts, slides, transparencies, postage and the like. Actual costs could range from about $5 per location for a simple session with a printed agenda to $50 or more per site for meetings featuring videotaped segments. Masters budgets a further $20 an hour per location for the call to the bridge and $20 an hour per location for toll charges.

In addition, large meetings require special speakerphone-type devices or amplifying equipment whose rental costs range from approximately $20 to $110 a month, Masters indicates. Staff time is another cost. Masters figures two hours to assemble and pack printed materials for mailing for every 20 to 30 sites.

The University of Texas Health Science Center regularly offers large-scale conferences. A typical program, according to Masters (formerly the director of the Center's Teleconference Network of Texas), offers a discussion of such subjects as hypertension to approximately 100 area hospitals through dedicated telephone lines. From five to 10 persons participate at each hospital. These sessions could be conducted using a "meet me" conference bridge. If they were, a typical session would include the items shown in Table 5.2.

In this instance, all additional costs associated with the program were borne by the hospitals receiving the transmission. These included the time of persons acting as site coordinators, and the supplying of the rooms, necessary projection equipment and refreshments.

One-way Video Conferences

By adding video, sponsors can greatly enhance the appeal of a teleconference. But they also significantly boost the costs. Video production itself is expensive, and meetings-by-screen also require far most costly transmission and receiving arrangements. Figure 5.1, a vendor's price quote checklist, illustrates some of the variables that must be considered when contemplating a video conference.

The bills for the three video conferences outlined in Tables 5.3, 5.4 and

Table 5.2   Cost of a Large Audio Conference

Preproduction	
Faculty honoraria	$ 200
On-site support (preparation and delivery to site of slides, transparencies and agendas)	2,000
Transmission	
Call to bridge	2,000
Fee for bridge	2,000
Transmission/reception equipment ($45/mo. per site spread over four sessions/mo.)	1,125
Total Cost	$7,325
Cost Per Person	$7-$15

Source: Jonna Lee Masters.

5.5—$45,400, $108,400 and $405,000—underscore that point. But it should be noted that the precise costs covered differ from example to example. Fees paid to outside vendors are generally monitored closely. But expenses for staff support or other services—ranging from time spent in program development to the cost of organizational telephone calls—are not accounted for uniformly. As a result, the budgets are not fully comparable.

First Boston Corp. Seminar

As a service to its clients, in early 1982 First Boston Corp. (New York, NY) broadcast a 2½-hour seminar explaining a new Securities and Exchange Commission regulation. According to Gregory Ferris, audiovisual manager, some 800 clients attended in 10 cities across the country: Atlanta, Boston, Chicago, Cleveland, Dallas, Houston, Los Angeles, Pittsburgh, San Francisco and New York, where the conference was produced. Satellite Networking Associates, which networked the meeting, used PBS stations in all but four locations. Table 5.3 shows the items included in the budget.

The First Boston meeting was not an example of a teleconference's being substituted for a face-to-face meeting: without teleconferencing, the meeting would never have been held. With just one day of the speakers' time available, the company could not have delivered the information face-to-face to the locations involved; therefore, the cost savings over traditional means is incalculable.

Figure 5.1 VideoNet Price Quote Checklist

GENERAL

1. Customer name _____
   Person to contact (phone) _____
2. Purpose of meeting (corporate, seminar, sales) _____
3. Date (day/week) _____
   Time of day (morning, noon, evening) _____
   Length of Program (live) _____
4. Amenable to existing systems such as PBS? _____
5. Question & Answer period between locations? _____
   How long? _____
6. Will fee be charged? Amount? _____ _____
7. One time event or possibly weekly/monthly? _____

ORIGINATING SITE

1. Where? (Hotel, Corporate HQ, Studio) _____
   City? _____
2. Speakers (corporate types, professional, etc.) _____
   Experience before cameras? _____
   Panel discussion/Single podium? _____
3. How extensive TV capabilities _____
   Number of cameras?, Scripting? _____
   Taping program? _____
   Teleprompter? _____
   Preparation of graphics? _____
   Staging, Full production? _____

RECEIVING SITES

Cities *must* receive _____

Cities *would like* to receive _____

Approximate number of people at each location _____

Desire for large screen TV (25 to 25,000 people)
vs. TV monitors _____

Desire for catering to be included in cost
estimate _____

GENERAL ADMINISTRATION

Desire/Need for . . . Preparation of brochures,
                     handouts, etc. _____

                     Preparation of pre-tapes to
                     introduce _____

Source: VideoNet. Reprinted with permission.

Table 5.3   First Boston Corp. Teleconference Budget

Origination	
Studio	
(including rehearsal)	$ 9,300
Uplink and backup and satellite time	
(2½-hour show, 1-hour testing)	4,500
Teleprompter	300
Art director/lighting director/	
makeup/crew	1,800
Site coordination, site surveys,	
downlinks, monitoring	10,600
Site surveys varied between $200	
and $500 per location	
Downlinking, including rental of space,	
TV monitors or projectors and	
telephones, varied between $1,300 and	
$1,600 per location	
Catering (estimated)	2,000
Printing (estimated)	1,000
Postage (estimated)	1,000
Local transportation (estimated)	1,500
Interactive telephone/	1,500
audio bridge	
Insurance	600
Satellite Networking Associates	7,600
mark-up	
Site survey of Philadelphia	
(location later cancelled)	500
New York City tax	3,200
Total Cost	$45,400
Cost Per Person	$ 57

Source: First Boston Corp.

Allied Van Lines Sales Conference

The experience of Allied Van Lines, by contrast, provides a clear index of the potential savings over conventional meetings. For its traditional series of regional sales meetings, Allied substituted a four-hour teleconference beamed from Chicago to 26 cities in spring 1981. Using Holiday Inn's HI-NET system sites for most locations, it reached 1300 agents nation-wide. The teleconference, organized by VideoNet, replaced a series of 2½-day meetings in six cities that reached only 650 agents in 1980.

By using teleconferencing, Allied spent less money—just $108,400—explains Bradford, compared with $173,150 to stage its traditional road show. And since the attendance at the 1981 session was double that in 1980, the per-person costs reflect an even more dramatic savings. Per-participant costs came to just $83, less than one-third the 1980 outlay. More people attended, Bradford explains, because the per-person costs were less. Not only were the meeting locations closer to the homes of many potential participants, but costs were not incurred for several days' lodging and meals for each participant.

A comparison of costs is given in Table 5.4.

Table 5.4   Comparison of 1980 and 1981 Allied Van Lines Sales Meeting

	1980 (Face-to-face meeting)	1981 (Tele-conference)
Employee expense		
(air fare, hotels, meals)	$ 54,000	$ 13,000
Facilities		
Meeting rooms/refreshments	44,000	—
Downlinks (including projection equipment, meeting rooms and refreshments)	—	42,000
Production		
Face-to-face meeting:		
Audiovisual show, including live music, projection equipment, production crew	89,000	—
Teleconference:		
Transmission	—	8,000
Trucks and facilities	—	12,000
Staging	—	2,000
Rehearsal	—	6,000
Video tape of previous year's meeting from slides and other elements	—	9,000
Coaching	—	15,000
Printing	1,100	300
Air freight	1,300	1,100
Gross Cost	$ 189,400	$108,400
Revenues		
($25 per person)	¹16,250	—
Net Cost	$ 173,150	$108,400
Cost Per Person	$266	$83

Source: Allied Van Lines.

A Major Pharmaceutical Company Teleconference

A major pharmaceutical company held an even larger teleconference, arranged by TeleConcepts in Communications, in fall 1981. Some 8000 physicians in 20 cities attended a three-hour session introducing a new antibiotic. While the cost of the session, $405,000, was almost four times that of the Allied meeting, the per-participant cost was significantly lower— just $51—because the cost was spread over a larger audience that assembled in hotels across the country. The budget for this conference is given in Table 5.5.

American Council on Education Teleconference

Not all institutions, of course, can afford such hefty out-of-pocket costs, even when the per-participant levy is low. But occasionally it is possible to cut costs by receiving services at a discount. The American Council on Education did just that in 1980, reports Daryl Ferguson, public affairs officer, with a two-hour teleconference on current issues in higher education beamed to more than 50 locations in 35 states. While the first hour was one-way video, one-way audio only, the second hour included a modified two-way audio facility.

Originating from station KUON-TV in Lincoln, NE, which substantially reduced its fees for the organization, the total cost came to just under $12,000. In addition, the Joint Council on Telecommunications donated consulting time valued at $2200. Out-of-pocket expenses included the items shown in Table 5.6.

Two-way Video Conferences

Primarily because of their high cost, two-way full-motion video conferences are by far the most rare. The additional production and transmission costs involved make sense when either the participation or facial reactions of conferencers is significant. Despite its high price, two-way video conferencing can be cost-effective if it eliminates the need for several people to travel a great distance.

At present, the major commercially available system for two-way, full-motion video conferencing is Picturephone® Meeting Service (PMS), inaugurated by AT&T in July 1982. This full-color system has been offered in two ways: users can rent public rooms or build rooms on their own premises. As of January 1983, PMS will be offered by American Bell, Inc. (AT&T's unregulated and independent subsidiary), while teleconferencing transmission facilities and consulting services will be handled by AT&T.

Table 5.5 Pharmaceutical Company Teleconference Budget

Origination

Studio	
(including 3 cameras, rehearsal	
day and air day)	$ 20,000
Visuals	10,000
Teleprompter and operator	
(2 rehearsal days and air day)	2,000
Lighting director	1,000
Director and producer	8,500
Associate director	1,000
Production assistant	1,000
Custom set	22,000

Transmission

Uplinks, satellite time and land lines to	
downtown hotels	2,600
Downlinks — TVROs	15,000
Telephone bank	
(3 telephones per site, 12 telephones at	
origination site, engineering and	
production coordinators)	15,000
Loops and connections	10,000
Long lines, interconnect and miscellaneous	
charges	12,600

Exhibition

Room rental	60,000
Large screen projectors	60,000
Engineering field coordinators, projection-	
ists, interactive transmission device	
and local audio arrangements	17,300

Network coordination, testing and

supervising, site surveys	7,000

Miscellaneous

(including estimated costs of program	
development and speech writing, publicity,	
catering, meeting coordinators)	$140,000
Total Cost	$405,000
Cost Per Person	$51

Source: TeleConcepts in Communications, Inc.

AT&T planned to make the service available in 16 cities in 1982: Atlanta, Boston, Buffalo, Chicago, Cincinnati, Cleveland, Columbus, Dallas, Detroit, Houston, Los Angeles, New York, Pittsburgh, Philadelphia, San Francisco and Washington, D.C. Public rooms were to be built in all ex-

Table 5.6   ACE Teleconference Budget

Staff travel	$ 1,570
(round-trip travel expenses for 5 ACE staff members to Lincoln)	
Meeting expenses	
Studio and production facilities	3,960
(including camera, video prompter, set design, makeup, producer, director, uplink, rehearsal)	
Video tape stock	400
Typing of teleprompter copy	3
Downlinks to 4 viewing centers	500
(remaining downlinks provided free of charge)	
Postage and mailing supplies	1,300
Printing	1,390
Data and word processing	1,010
Telephone calls	840
Coordination fees	920
(to Public Service Satellite Consortium)	
Miscellaneous	30
Total Cost	$11,923

Source: American Council on Education.

cept Buffalo, Cincinnati, Cleveland and Columbus during 1982. In 1983, AT&T plans to extend service to 26 additional cities: Albany, Charlotte, Denver, Des Moines, Greensboro, Hartford, Indianapolis, Kansas City (MO), Louisville, Memphis, Miami, Milwaukee, Minneapolis, Nashville, Omaha, Orlando, Phoenix, Raleigh, Richmond, Rochester, St. Louis, San Antonio, San Diego, Seattle, Syracuse and Tulsa.

As of May 1982 a one-hour meeting held in public conference rooms in New York and Washington typically would have cost $1340. A similar session held between New York and Los Angeles would come to $2380.

For customers installing their own rooms, the tab would be lower: just $600 for a one-hour New York-to-Washington conference and $1640 for a one-hour New York-to-Los Angeles session. But the fees for installing on-premises facilities are steep. Customers installing a typical room will pay one-time Bell System installation charges of $117,500, as well as monthly equipment rentals of $11,760. In addition, there is a monthly charge to connect each room to the nearest Bell System facility.

Satellite Business Systems (SBS) also provides a permanent two-way video conferencing capability through its Communications Network Service. However, SBS does not promote the teleconferencing service alone, but as part of its integrated office communications package.

If AT&T's facilities are not available at needed locations, a temporary network can be assembled. According to the Public Service Satellite Consortium (PSSC), the cost of a two-hour *ad hoc* video conference beween two major East Coast and West Coast cities could range from $14,000 to $21,000. That sum reflects a simple production with two cameras, no special set design and no preproduction. A more elaborate conference would, of course, increase the price. There are several companies that are involved in two-way video conferencing, primarily on an *ad hoc* basis and these are listed in the Appendix at the end of this book.

INTERNATIONAL TELECONFERENCING

Just as the cost of travel to Europe is far more than a trip to a nearby city, the cost of teleconferencing is multiplied when transmitting overseas. In an audio teleconference using commercially available bridging equipment, the major additional factor is the added cost of dialing overseas. But transmitting video material overseas is more complicated.

Permanent International Service

At least one permanent commercial service is under development. In March 1982, Intercontinental Hotels Corp. and COMSAT General Corp. announced a joint venture to provide the first international two-way video conferencing system available to the public. Scheduled to begin in early 1983, the service initially will link New York and London. Additional locations are under consideration for such cities as Houston, Paris and Toyko.

The system is designed for small executive conferences but can be adapted for presentations to a maximum of 12 persons. Services offered will include document transmission, freeze-frame video and full-motion video, and costs will vary according to the services selected. A typical freeze-frame video meeting is estimated to cost between $1500 and $2000 an hour.

Ad Hoc International Services

An *ad hoc* overseas segment can run considerably higher. According to Bob Bossi, manager of television services for International Telephone & Telegraph Corp., a rough estimate for the cost of overseas transmission is approximately $7000 an hour, although it could be less. While each country has a different fee structure, says Badoud of VideoNet, planners should add approximately $20,000 for each site desired in Europe or South America. Canadian costs are somewhat lower.

Several factors go into that estimate. International charges for uplinks and downlinks are higher than the U.S. equivalents. Since items such as video projectors are not as available, equipment in foreign locations generally costs more. In addition, special equipment is needed at the origination site to integrate the signals.

To extend a preexisting U.S. teleconference to an audience of 50 in London in spring 1982, Badoud calculates, would have cost $14,480 for one hour and $21,435 for three hours. This estimate includes projection equipment but does not reflect the cost of catering or room rental. To beam the same teleconference to Sao Paulo, Brazil, fees would be $14,420 for one hour and $21,635 for three hours. Sending the same teleconference to both countries simultaneously would cost nearly as much as holding two separate teleconferences. To send the same teleconference to both London and Sao Paulo, says Badoud, would come to $25,900 or so for one hour and nearly $40,000 for three hours.

HIDDEN COSTS OF TELECONFERENCING

Much of the previous discussion has concerned direct out-of-pocket costs of *ad hoc* teleconferencing. But there are other costs, including the time and internal resources expended on activities such as letter-writing and duplication, that are rarely budgeted, notes TeleConcepts' Johnson.

In-house Time

The most significant of these is in-house time. In some cases, it takes the same amount of time to organize a teleconference as it does to hold a face-to-face meeting. Allied's Bradford states, for example, that the same amount of staff time—including program development, speech writing and the like—was involved in arranging both the company's 1980 face-to-face sessions and its 1981 teleconference.

In other instances, the time invested can be much greater than that for a conventional meeting, and overlooking such costs can be misleading. At First Boston, for instance, Ferris spent the better part of ten days working on the teleconference. He calculates the cost of his time, as well as that of others who participated in the preparation, to be worth at least $5000. ACE's Ferguson estimates the value of the staff time invested in preparing the video program and traveling to Nebraska at $3950.

Making Arrangements

Such intangibles are not the only hidden costs. The elaborate arrangements needed to coordinate an event that takes place at many sites often spark unexpected costs. Telephone bills can soar. And often users will consider it advisable, as Allied did, to hold at least one in-person meeting for all local coordinators; an effort that cost the company $13,000.

Costs can add up in other areas as well. When an organization is making a sizeable investment in a teleconference, some experts suggest, it is likely to spend more on mundane details as well—putting the agenda and other supporting materials in a plastic loose-leaf notebook, for example, rather than a manila envelope.

COST OF BUILDING AN IN-HOUSE SYSTEM

An organization involved with extensive teleconferencing, whether audio or video, may be wise to consider installing an internal conferencing capability. Precisely when such a move is warranted depends on an organization's needs.

As of spring 1982, estimates the senior audiovisual officer of a major corporation, hundreds of companies had developed their own dedicated audio systems, while those with the far more costly two-way full-motion video operations numbered less than a dozen.

Audio Conference Installations

Outfitting an audio conference room is the less expensive option. Susan Pereyra of Connex International estimates that the cost of equipping a room for audio conferencing can range from $2000 for a simple format to $50,000 or more.

Honeywell, for example, operated six audio conference rooms in the third quarter of 1981. The most recent had been built for approximately $30,000, says Rick Whiting. That estimate included the cost of furniture and projection equipment. In addition, the company leased an audio terminal and facsimile machine for approximately $550 a month and maintained a clerical coordinator to handle room booking and scheduling.

Video Conference Installations

Companies can also design and install video systems on their own. While the cost of basic equipment such as cameras and monitors remains constant, costs of video processing equipment and the transmission itself vary widely. In essence, explains Greg Paulsen, director, consulting services and training division for VideoNet, analog systems have a low equipment cost but high monthly transmission cost. Digital systems, by contrast, entail a high capital investment but have relatively inexpensive transmission costs.

For full-motion video, the initial cost of a system using analog transmission ranges from $25,000 for a simple conference room to as much as $350,000 a room or more. In addition, notes Paulsen, one-way analog transmission costs for a dedicated circuit connection (24 hours a day, seven days a week) would come to an average $1.2 million a year. A two-way analog transmission, of course, would be twice that rate.

On the other hand, building a system based on digital transmission can run from $175,000 a room to $500,000 or more. But a two-way digital transmission path could come to as little as $15,000 a month.

Once a network is built, Paulsen suggests, organizations should budget roughly 1% of purchase price per year for maintenance and one person per location to handle room bookings.

CONTROLLING COSTS

Adequate Planning

The key to controlling teleconferencing costs is careful planning. As with any other meeting, the better planned the session, the lower the likelihood of unexpected or last-minute bills.

This is especially true with video conferences. Because the cost of each component is high, a decision to add, for example, an extra site or two will boost the budget substantially. Changes in existing conference elements can be surprisingly expensive as well. And the later the shift, the higher the tab.

"In television, time is money," notes Adele Brown. For example, the fee for preparing visuals on short notice could be double the normal rate, she says. Similarly, observes TeleConcepts' Johnson, an organization could be forced to use more expensive equipment if the preferred materials are no longer available. Even the tiniest of changes can prove costly. If a location is switched at the last minute, just sending a program correction to 20 sites by express mail could add a quick $200 to the final bill.

Comparison Shopping

Another good way to hold down costs is by shopping around. Allied's Bradford suggests soliciting bids from at least three vendors. She herself received bids ranging from $74,400 to $170,000 for Allied's 1982 video teleconference. In dealing with vendors, advises Daryl Ferguson, nonprofit groups particularly should use whatever leverage they can muster to line up discounts.

Another critical financial issue turns on the degree of redundancy in the final equipment configuration. A user can cut costs by limiting the amount of back-up equipment or testing time. But such a decision should be weighed carefully. One teleconference vendor indicated that the failure rate experienced during testing had run as high as two instances in 100. In fact, in comparing vendors' estimates, a user should determine what, if any, insurance is provided against the loss of all or part of a transmission, and under what circumstances a refund would be issued.

Using Outside Consultants

Unless it boasts a strong video staff, many users suggest that an organization engage an outside consultant for at least its first teleconference. Organizations can pare costs by making their own arrangements, some vendors say, only if they have individuals experienced enough to put together and cost out a session properly. In all instances, both users and vendors must fully understand what services will be provided.

CONCLUSIONS

The costs of holding a teleconference vary dramatically. But when wisely applied, this technology holds the promise of trimming the costs of a comparable face-to-face meeting.

While much discussion focuses on total meeting costs, a compelling gauge of the cost-effectiveness of a teleconference is the per-person-per-hour charge. Because many of the costs of a teleconference are one-time fixed costs, the per-person tab drops markedly as the size of the audience increases. In assessing the financial benefits of a teleconference, potential users must focus on per-person costs as well as the total bill.

On the other hand, the adoption of teleconferencing does not automatically mean that less travel will occur. Teleconference users may feel an increased need for communication, Honeywell's Whiting believes. Furthermore, he suggests, teleconferencing may spotlight new reasons, such as new sales opportunities, for travel. Whiting notes that most organizations have less money available in the budget than reasons for trips. As a result, he speculates, the money not spent on one teleconference may not be money saved; it may just be spent on a trip or teleconference that would otherwise not have been funded. To save money in the long run through teleconferencing, Whiting contends, organizations must closely manage both their teleconferencing and travel budgets.

6

Anatomy of a Teleconference

by James W. Johnson

INTRODUCTION

The gestation of a teleconference can vary from a few days to a year or more. During that time, from the first ideas, to its birth, growth and finally maturation at "air" time, a lot happens. In order to examine the anatomy of a teleconference we will start at this first "idea" level.

The teleconference is first and foremost a communications medium—thus, the need to communicate must be the father of that first idea. Of course, we know that communication can take place on many levels; from the simplest visual actions—a look, a smile, a frown—to a few printed words, or sounds, to the most complex visual or aural messages. Before we can select an appropriate level, however, our communications needs must be identified and then examined in the light of my "First Law of Communications":

Any communication is maximized at its simplest effective level.

Assuming our wish to maximize, we examine the possibilities starting at the simplest level of communication and stopping when we reach the point of effectiveness. Will a word or a look do? Need it be a letter, or a memo? Maybe a few spoken words will work, or a phone call. Try the simplest medium first—and move upward with reluctance. Examine each possibility thoroughly before you move on. And don't forget to identify the true message. Actual words or actions may not be what they seem. An inter-office memo may really be saying: "I'm the boss here, and I'm pretty darn important, so you'd better listen when I speak." The actual words in the memo may be inconsequential.

125

Ask yourself: What is the message? What media can deliver it? What media can contribute to the message? Can any medium actually *be* the message? The answers to these questions will help us find the most effective medium. Refer to Figure 6.1, my "Square of Intrinsic Effectiveness." We see that the simpler the medium that can deliver the more complex a message, the greater the intrinsic effectiveness.

ANALYZING THE NEED FOR A TELECONFERENCE

The reasons for using a teleconference can range from the need to impart an important, complex message to a widely located audience, to the need to establish credibility with shareholders, instill enthusiasm in a sales force or perhaps to sell a product.

Figure 6.1 Square of Intrinsic Effectiveness

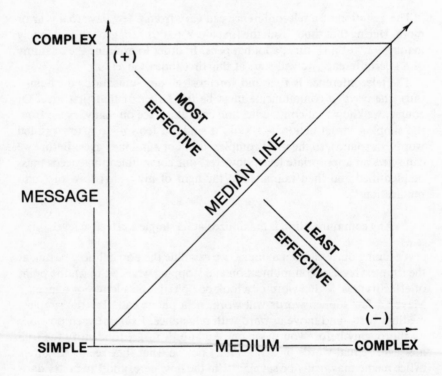

The decision to consider a teleconference should be made after rejecting simpler communications forms. Then comes selection of the most effective teleconference mode: audio conference; locally enhanced audio conference (slides, films, video tapes, transparencies, etc.); remotely enhanced audio conference (slow-scan, electronic blackboard, limited video, etc.); full-motion (one-way) video/two-way audio conference; two-way video/two-way audio conference.* Computer assisted message interchange (CAMI) should also be considered. Known as computer conferencing, this is not strictly teleconferencing since it is not a "live" (real-time) event, but it can be effective and inexpensive when real-time impact is unnecessary.

Even accompanying printed material is part of the conferencing process. Invitations, bulletin board announcements, newspaper ads, programs, product information and press kits, monographs, newsletters—all can be part of the teleconference medium. And while all add complexity, not all add effectiveness. Figure 6.2 suggests a format for analyzing a communications need. Determine where each teleconferencing mode falls in Figure 6.1, The Square of Intrinsic Effectiveness. Note which modes fall into the "plus" side, and which into the "minus" side. The simplest "plus" will indicate the teleconference mode that best meets your communications need. If there are no "pluses," a teleconference is probably the wrong medium.

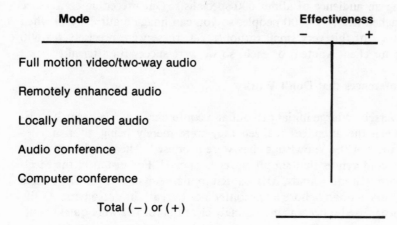

Figure 6.2 Communications Needs Analysis

Mode	Effectiveness −	+
Full motion video/two-way audio		
Remotely enhanced audio		
Locally enhanced audio		
Audio conference		
Computer conference		
Total (−) or (+)		

---

*This is a unique form of teleconferencing, usually limited to two sites, and the reader is referred to Chapters 2 and 8 for more information on two-way video/two-way audio conferencing.

WHAT TELECONFERENCES CAN ACCOMPLISH

I was once asked what was the worst disaster that could befall a teleconference. The interviewer expected me to say something about falling satellites or lost pictures, but my answer was that the users' expectations are not fulfilled. While this holds true for many things, the teleconference is particularly sensitive to this shortcoming because of its novelty and because it is a live event—there are no second "takes."

Teleconferences are relatively new and most users do not have a clear vision of a conference and its benefits. The teleconference comes in many forms and is poorly understood. For example, let's say you do a teleconference to market a medicine. What are your goals? A large audience? A special audience? A grateful audience who will remember and prescribe your product? Is success to be measured in increased sales of the product, or by heightened awareness of your company and growth of all your sales? Is it to expand a market, create a new market, enlarge your share of a market, reposition your product within a market? Is real success your own promotion to a better job, or the request for future teleconferences? What kind of comments do you hope for from your audience? Try to answer these questions, and pose still others of your own. Decide beforehand what the criteria for success will be, and then stick to them. Don't forget value. A large audience obtained at an exorbitant price may not be a success. An ineffective conference that misses its mark is no bargain at any price. We once produced a six-city teleconference on a very esoteric medical subject, reaching an audience of about 600 specialists. Our preceding conference had reached nearly 13,000 people, so you can imagine our chagrin when we learned of this very small turnout. Yet, these were precisely *the* 600 people our client wanted to reach, so we were successful after all!

Teleconferences that Don't Work

Conversely, a large initial turnout at a conference quickly became a disaster when the attendees realized they were merely being given a sales pitch, and not the knowledge they were promised. Then, there are some notable conferences that simply never happened. For instance, the Steelcase Corp. (Grand Rapids, MI) wanted to market its new modular offices a few years ago, and chose a teleconference to reach its customers. As the plans progressed it became increasingly clear that orders were quickly outdistancing supply, so that the reason for the teleconference simply disappeared. With only a few weeks to go, the teleconference was cancelled. Xerox Corp., in a similar manner, cancelled an ambitious international telecast at the last minute because corporate events changed the nature of the message. Xerox had enough sense to cancel, at a substantial cost, rather than go ahead when it saw that the medium no longer matched the need.

When Teleconferencing is the Way to Go

Once you've determined that a teleconference has the best intrinsic effectiveness, the next step is to determine which type of teleconference best suits your needs. We have found it helpful to assign a "televalue" to each project; i.e., to rate each of the different elements involved on a scale of one to five. Figure 6.3 suggests a method for conducting this evaluation, and gives two examples for which values have been determined.

In example A, an announcement to staff, the televalue is a low 16; in example B, a press announcement, the televalue is a maximum of 40. These televalues can be checked against the list of values given for the various teleconference modes given in Figure 6.4. The idea is to seek a "televalue" match between the communications need (determined in Figure 6.3) and the teleconferencing mode (see Figure 6.4). (While this produces no hard and fast decision, it can expedite the decision-making process by quantifying some of the factors involved.)

Now that we've talked about teleconferencing as a whole, let's discuss some of its parts.

ANATOMY OF A VIDEO CONFERENCE

All teleconferences require planning: the simpler modes, such as audio conferences, require less, while full-motion video requires more. The rest of this chapter will address the interactive video conference, since it contains all the elements of the simpler forms of teleconferencing. Each element of a video conference will be considered in turn.

Origination

This pertains to all the parts that go into the program proper—i.e., the televised event, in the case of a video conference. It starts with planning sessions. What are we trying to say? How do we wish to say it? If we're addressing an outside audience, we must plan with great care. Who will participate? Who will write the program? Where will this writer get his or her data and direction? A word about this: a staff writer is probably not qualified to write an effective teleconference, because it is a distinct media form. For this reason alone, the teleconference planner is well advised to obtain the services of a specialist writer, rather than put an unfair burden on a staff writer placed in a new and demanding environment.

Figure 6.3 Evaluating Need for a Teleconference

Examine the "telefactor" listed below, and assign a value of from 1 to 5 to each in light of your communication requirements. Check your televalue with the chart given in Figure 6.4.

		Value
1. Number of participants	1 = few 5 = many	_____
2. Need for personal impact	1 = low 5 = high	_____
3. Familiarity with viewer	1 = friends 5 = strangers	_____
4. Visual complexity	1 = little 5 = much	_____
5. Available funds	1 = low 5 = high	_____
6. Distance between meeting points	1 = short 5 = long	_____
7. Need to visualize	1 = little 5 = much	_____
8. Need for interaction	1 = none 5 = much	_____
	Total (equals televalue)	_____

If your communications need is to announce sales figures to your staff, then your score might look like column A, below. If it is to announce a medical breakthrough to doctors and the press, then it might look like column B.

	A: Staff announcement	B: Major press conference
1. Number of participants	2	5
2. Need for personal impact	2	5
3. Familiarity with viewer	1	5
4. Visual complexity	2	5
5. Available funds	1	5
6. Distance between meeting points	3	5
7. Need to visualize	2	5
8. Need for interaction	3	5
Total	16	40

First Person or Third

The camera eye can be first or third person—or a combination of both. An early choice is necessary, however, in order to structure the program effectively. A first person camera represents each individual member of the audience. The speaker looks right into the camera eye—into *your* eye. The speaker's message is personal, and when used well, powerful and ef-

Figure 6.4 "Values" of Teleconferencing Modes

Modes	Televalue
Full motion video/two-way audio	40
Remotely enhanced audio	30
Locally enhanced audio	20
Audio conference	10
Computer conference	5

fective. A direct sales message is maximized by a first person camera approach. The viewer is targeted carefully and must squirm to avoid the message. On the other hand, there are times when a soft sell is required— where the viewer is more comfortable as a third person "watcher" instead of an active participant. This calls for strong interaction among the participants, with audience involvement restricted to questions and answers (Q and A). It also calls for clearly written dialogue and subtle attention to marketing goals.

Environment

Environment is another early consideration. What and where should the setting be? Here's another powerful application of "the medium is the message." Originating from a factory or a headquarters building can have meaningful implications for the viewer. For example, New York City sends a different message than, say, Roanoke, VA. But different is not necessarily best. Choose the site that best supports the message of your program. This applies to the choice of "studio" (boardroom, factory, permanent TV facility, etc.) as well. Remember that control is the essence of success. This means that when it comes to ease and proficiency of production a TV studio is superior to a hotel ballroom, which in turn is superior to a company cafeteria. While a studio can be created almost anywhere, opt for a permanent facility whenever possible.

After selecting the environment for your production, consider the environment for your presenters: the stage "set." Again there is a message in the medium. The appearance of the set speaks strongly for both the sponsor and its message. Simplicity, good taste, quality, appropriateness, effectiveness, beauty, apparent cost and size are all qualities that may describe the set and client as well. For example, a national emergency requires a very businesslike, no-frills setting: a lectern with draped flags, perhaps. The introduction of a new consumer product might be done in a realistic setting—a shop, or kitchen. A press conference should be sterile—the setting must not detract from the message. A sales meeting might be modern, gimmicky, slick, exciting. And a serious medical symposium should have a businesslike and perhaps scientific setting.

A professional scenic artist might not be required if you're speaking to your employees. One is definitely needed however, if you wish to sell a product or motivate strangers. The professional knows what works well for the cameras, what will make the speakers easy to see, which colors and textures will be effective and which will not. And he uses all this to achieve the goal-oriented environment you need.

By the way, it's a good idea to design a setting that will work for you more than once. The "X-Y-Z Conference Center" will become recognizable as your company's "signature" on its conferences, and instill a strong feeling of continuity. It will also save a lot of money over the course of several teleconferences.

There is a trend toward the "no-set" set. But remember that whatever your camera sees behind your speaker is, indeed, a set, whether or not it's designed. Unfortunately, most natural surroundings contribute negatively to the program: people blend into paneled walls, have pictures or plants seemingly growing out of their heads, share the scene with distracting window blinds, or turn green to match the background. This is not necessary. Dozens of production manuals offer excellent advice for low-budget programs, and most scenic artists will work within your budget, even if it's modest.

Lighting

The lighting of a telecast is largely misunderstood, and respected for the wrong reasons. Most people think of stage lights as merely brighter versions of existing light. Not so. The differentiating factor lies in *control.* Stage lights (TV lights) are unique in their ability to control. They can be highly directional, very selective in their area, variable in brightness, and, through the use of color, creators of mood and style.

The television camera does not see as you do. While it can discern a wider range of light gradations than film, it requires more light to function properly than the human eye does. Thus a specialist is needed to position lights effectively to create a mood, flatter the speakers, eliminate strange shadows—in short, create the illusion that there is no special light at all. If you can afford a lighting director, it's a valuable investment. Good lighting can make a medium-quality camera look better, but the best camera in the world cannot compensate for poor lighting. A lighting director who is familiar with teleconferencing's requirements will understand the TV projection system's strengths and weaknesses, and have the ability to light for live, multi-camera production.

Makeup

Makeup is always required. TV is seldom kind to a person's appearance. People perspire and start to shine, and have wrinkles, lines, bags, bald spots, invisible eyebrows, jowls and other blemishes which can be helped by makeup.

A professional artist is essential for a professional production. A knowledge of modeling and color, an appreciation of how to compensate for TV's largely overhead lighting, and an ability to work quickly and well with sometimes nervous performers are invaluable assets. Professionals will never object to being made-up. Some nonprofessionals may resist, however. My experience with this has taught me never to ask people if they want makeup, but rather just routinely assume that they will be made-up.

Studio Facilities

Next, we come to the studio facilities. This is not the place to list the many brands and qualities of equipment available today: suffice it to say that "more expensive" is usually better and, as always, your choice should reflect your goals.

Cameras and Recorders

For example, how many cameras are required? The simplest of pick-ups requires two, even if one is merely a backup for the other. Even a highly complex production rarely requires more than four cameras. Three is most common. This allows two differing shots of the speaker and another of the listeners, with a chance to cut away to an easel shot of a visual element once in a while.

Even though a teleconference is live, we always record it. At the simplest level, an inexpensive video cassette should be made for archival purposes. At the other end of the scale, a 1-inch or 2-inch broadcast-quality recording can be edited and used for future viewings—it can even be transferred satisfactorily to 16mm color film. Subject matter and ancillary uses will determine the level of recording quality you need. (Sometimes it's a good idea to isolate one or more camera outputs and record them separately. Later on, when editing the telecast, these secondary recordings can be an invaluable source of cut-away, or alternate, shots.)

Rehearsal

Every person important enough to "star" in a teleconference is important enough to place a very high value on personal time, and may resent taking the time to rehearse. Unless you are a strong producer, an adversary relationship may develop, and both you and the performers will pay the price in a poor presentation. Rehearsal is an absolute must if a script with visuals is part of the program.

Otherwise intelligent individuals, who work hard and diligently practice at other endeavors, assume they can just walk on like Johnny Carson and "wing it." What they don't realize is that Johnny Carson doesn't *ever* wing it; he rehearses, and he's a consummate pro. So don't accept any excuses. Build in the time requirement right at the start. If your speaker won't give enough time, postpone the conference to a date when he *does* have time, or don't do it at all.

How much time is enough? There must be several short (two-hour) meetings early on to go over the speaker's ideas and to get his input. Then he must review and approve the script and visuals *promptly*. After that, you should have a separate prompter rehearsal, followed by one or two camera rehearsals and then a run-through of the entire program. By the time everyone has been through the teleconference several times, they are relaxed and are using the prompter easily and well.

Scripts and Prompters

Not all programs require a script, therefore not all programs require prompting. However, when exact words are the order of the day, prompters are essential. Almost every studio worth its salt offers prompters or can rent some for your program. Of course, short segments can always be done with carefully lettered cue cards (idiot cards), but one must look slightly off-camera in order to read them, and eye contact is lost. Choice of prompter operator is also critical. Seek the most experienced you can find. It looks easy, but so does golf—until you try it.

Much depends on the kind of telecast you are presenting. Simple information exchange should be done extemporaneously, using a general outline for a guide. A marketing event, where content is king, should be closely scripted and prompted. It is desirable to have the script committed to prompter rolls well in advance of the event, for review by the participants.

Two or three days before the conference, we like to bring the prompting equipment to the office or hotel suite where the "cast" is gathered, and coach each speaker through his or her part, making minor changes for the sake of speech comfort. Even if the speaker has approved his script—or has even written it himself—it sounds different when it is spoken. So from time to time we hear, "I wouldn't say it just like that," and we make numerous tactical, but nonsubstantive, changes to enable speakers to concentrate on the thoughts instead of the words.

Many speakers are accustomed to addressing a live audience, but television is an intimate medium. When President Reagan addresses the nation, he speaks to each of us personally. Many people believe that a speech can be televised and be effective TV without adaptation to the medium. It cannot. Effective teleconferencing demands a relaxed, informal, highly personal style. The audience wants to be spoken to or with, not at. The reason for so many dry, painfully boring teleconferences is the lack of understanding of this quality by the writer and/or the producer.

Use of Talent

On-camera personnel need a little thought too. Much of the time the "performers" are chosen because of their hierarchical positions and specialized knowledge. However, occasionally it is required that someone portray another person—a fellow employee, a patient, a doctor, an executive, etc. The temptation is great to use a *real* employee, patient, doctor, etc., for the part. Resist! However willing and well-meaning these real-life portrayers may be, they lack the understanding of the medium and the discipline that a professional actor possesses. Amateur talent may be inexpensive in the short run, but it exacts a high price in extra time required for shooting and reduced program quality. An amateur actor might be a reasonable low-budget solution to this problem.

Transmission

The origination is really just a television signal which must be delivered to be used. This is the transmission part of a teleconference, and involves different delivery modes combined in a network. Although it is wise to use the services of a professional vendor for transmission, it is helpful to know a few basics.

A TV signal can go over wires, through the air or through space via satellite. Equipment is sometimes already in place to achieve this. Other times it must be provided for your use. To transmit your program from a studio or other origination site to other locations you can use Telco (the local phone company, usually) which can send your program to a permanent uplink facility (satellite transmitting station). Or, you can bring in a portable uplink earth station (satellite dish) which will transmit the signal directly from your site to a communications satellite. (This is sometimes prohibitively expensive.) You might choose not to use a satellite at all, and instead have AT&T send it via IXC (interconnect) longlines (sometimes called "land lines") from city to city. This is generally more expensive (it can include unanticipated "special construction" charges), but can be more reliable under certain circumstances.

TVROs (receive-only downlink earth stations) pick up the signal transmitted by the satellite and make it available to the viewer. There are thousands of permanent TVROs and hundreds more mobile TVROs which can be rented and brought to any site for a teleconference. Thus Telco, AT&T, satellite transmissions or a combination of these will bring your signal along its way. Cost items (local loops, connections, etc.) are discussed in Chapters 2 and 5.

Interaction and coordination circuitry are also elements of transmission. Depending upon the complexity of the project, from one to three business phones are necessary at each receiving site. One is used for interactive audio, the others to coordinate engineering and programming events. Often one of the coordinating phones is equipped with an output jack so that, by plugging in the public address system, backup program audio can be transmitted over it in case of a transmission loss. A number of additional phones are necessary at the origination site as well for network, program and interaction coordination.

Backup Procedures

This is a good place to discuss backup procedures. Ideally, every aspect of a teleconference should have a backup. Since a teleconference is live, there are no second chances. We therefore spend a lot of time trying to imagine what can go wrong and what we'll do if it does. A corollary of Murphy's Law—"if anything can go wrong it will"—is Murphy's Anti-Law which says that anything you've backed up will never fail. So we rarely need to use a backup, as long as we have one!

What to back up? Well, we have made arrangements for a spare prompter roll, spare video tape for program inserts, spare slides (a complete set), spare cameras, spare tape recorders—even spare people. (What if your program moderator becomes ill the day of the telecast?) We have a spare transmission path, spare satellite dish circuitry and a spare transponder. Sometimes the program is even sent over two separate satellites. Backup projectors are routine, as are spare public address systems and spare phones—a necessity, especially if a foreign country is involved.

Our law of redundancy (backup) is "if you think of it, do it." We start by going over every inch, every smallest step, of every event. If Harry must be somewhere at 7 a.m., and we wonder if Harry might oversleep, we promptly make arrangements to wake him up, even if he's never been late for anything in his life. This principle applies at all times and in all circumstances. It is an antidote to Mr. Murphy's worst.

Exhibition

We call the next stage exhibition because it includes pictorial as well as auditory reproduction.

Monitors and Projectors

The TV image can be seen on a 19-inch monitor or receiver if the audience is small—under 15 persons. Larger audiences, depending on size, require a larger screen projector. (Actual sizes are usually limited by the ceiling height. The bottom of the screen should be at least four feet from the floor so the heads of the seated audience don't block the view. An 11-foot ceiling limits you to a 7x10-foot image, therefore, regardless of the size of your group.)

TV receivers are what you watch at home. Monitors are similar, except that they require a special kind of signal, are high quality and cost more. Either is suitable for teleconferencing. The smaller projectors (with an image of up to 4x6 feet) are not very bright, and are not suitable for very large audiences. Brighter (larger) projectors are of several different kinds. The brightest and sharpest of all projectors is also the largest, most difficult to install and operate, and the most expensive by several orders of magnitude. It is therefore reserved for very special applications and rarely used for teleconferencing. There are a number of medium-sized units, however, that are well-suited to video conference needs, and your producer can suggest these.

Survey the Meeting Site

There is a human element in exhibition. It starts with the need to survey each meeting site. You never know what you will run up against, and the day of the conference is usually too late to correct anything. Even if you have been to the site before, it's a good idea to take another look. Once we surveyed the grand ballroom of a Boston hotel rather too far in advance, and when we came to use it for a TV pick-up discovered it had been redecorated with floor-to-ceiling mirrors! We did a lot of fast (and expensive) emergency draping, to eliminate absolutely horrendous reflections.

When your network involves many cities, your vendor should supply an engineering field coordinator at each site to survey it, oversee the installation and operation of equipment, and take emergency action if required. He will also act as a liaison with the site representative, the client and the local phone company, and will instruct the local chairperson in the use of the interactive Q/A system.

Interaction

This brings us to the last essential element of a teleconference—interaction. Sometimes referred to as the Q/A, or interactive Q/A, it implies two-way communication. The simplest effective means of accomplishing this is usually by the use of audio only. There is rarely a need for the questioner at the receiving site to be seen by the expert at the origination location: 95% of today's teleconferences have interactive audio only. On the other hand, not all interaction consists of questions and answers. Decide who must be seen to be effective. Speakers of equal importance may deserve equal visualization.

While program sound is transmitted with the picture and is of high quality, return interactive sound is customarily sent via telephone, and provides relatively poor sound quality. This is not necessarily bad, however. Once, I offered to provide high-quality interactive sound at very little extra expense, making the questioner sound as good as the moderator in the studio. The response from my client was a vigorous "Are you kidding?" He properly understood something I had failed to grasp: the poorer sound quality established the remoteness of the meeting site and suggested the size of the network better than any word or picture could do. I never made that suggestion again.

Structuring an interactive Q/A session is important. The simplest and least controlled method consists of a member of the audience going to a microphone and posing an unscreened question. The question is heard by the audience and the moderator and answered by the moderator or his panel. The next level of control consists in having the audience write questions on cards. The cards are culled by the local moderator, and the writer of the question chosen is paged, comes to a microphone and asks his question on cue. At the next level, the moderator voices the question, thus assuming subject control. A still higher level of control is achieved when the local moderators at each site call a national question moderator at a central question control site. Their questions are culled by the national moderator, who determines both geographic and subject order. This information is then relayed to the moderators at the local sites and to the TV control room so a city identification image can be aired. A sensitive marketing event needs this high level of control. Some teleconferences take call-in questions in written form only. Both questions and answers are voiced by the presenter. This is not true interactivity, and calls into question the need for a teleconference at all.

TELECONFERENCING: THE WHOLE

Having examined the parts of a teleconference, let's look at the whole "animal." We have seen that each main part of a teleconference—program, origination, transmission, exhibition—can have many variations, depending on needs. The key to structuring your teleconference properly lies in selecting those elements that result in the most effective whole.

The chart given in Figure 6.5 lists the various teleconferencing parts and options discussed. By checking off your options, you will develop a profile that will help you to identify goals, expectations and ancillary requirements. The more complex your profile, the more planning, the more

Figure 6.5 Teleconference Profile Chart

Program	Simple	Average	Complex
1. Number of origination sites	☐ One	☐ Two	☐ More
2. Length of program	☐ Under 1 hour	☐ Under 3 hours	☐ Over 3 hours
3. Number of participants	☐ One or two	☐ 3 to 5	☐ Six or more
4. Script	☐ None	☐ Partial	☐ Complete
5. Interaction control	☐ None	☐ Some	☐ Much
6. Purpose	☐ Informational	☐ Mixed	☐ Marketing

Origination			
7. Cameras	☐ One	☐ Two	☐ Three or more
8. Recording	☐ None	☐ Archival	☐ Editing master
9. Crew members	☐ Under 5	☐ 5 to 10	☐ Over 10
10. Setting	☐ Drape/lectern	☐ Stock units	☐ Custom designed
11. Prompting	☐ None	☐ Camera mount	☐ Multiple
12. Lighting direction	☐ None	☐ Local	☐ Professional

Transmission			
13. Outgoing signal	☐ Fixed uplink	☐ Telco	☐ Mobile uplink
14. Network elements	☐ Permanent	☐ Mixed	☐ Custom
15. Reception	☐ Permanent TVRO	☐ Mobile TVRO	☐ Telco
16. Network size	☐ 2 to 10	☐ 11 to 25	☐ 26 or more
17. Number of phones per site	☐ 1	☐ 2	☐ 3

Exhibition			
18. Audience sizes	☐ Under 50	☐ 50 to 200	☐ Over 200
19. Room style	☐ Conference	☐ Hotel meeting	☐ Hotel ballroom
20. Display screen	☐ Monitor/ receiver	☐ Small projector	☐ Large projector

cost and the greater the level of expectation involved. And, the more complex your teleconference, the more you will need outside professional help to produce your event.

For example, a recent telecast done for the Johnson & Johnson Orthopaedic Division introduced a new hip prosthesis to an audience of physicians assembled in 35 sites throughout the country. When profiled (using the chart in Figure 6.5) this teleconference had a "complex" profile. Therefore it required considerable advance planning (nearly six months) and was most dependent upon the services of a professional producer, even though the company has an excellent in-house facility. On the other hand, a teleconference produced for Applied Systems Inc. (ASI) involved only one site (a hotel ballroom in Chicago and a guest speaker in Los Angeles), and had a "simple to average" profile. While this teleconference was also professionally produced, much of the production responsibility was assumed by ASI staff. In both cases, early profile analysis helped to assign responsibilities and to determine specifics from the start of planning. This was an effective way of maximizing program content and minimizing costs.

CONCLUSIONS

Teleconference tools are proliferating as the medium grows. New applications surface almost daily. The medium is catching the imagination of large segments of the population, and forcing it to grow in ever new and exciting directions. We should savor these moments of growth, for it seems that teleconferencing is experiencing a type of speeded-up evolution, where the species is growing new parts and adapting to new environments at a mind-reeling rate. By analyzing the needs this medium addresses, and by finding new applications for it, we can help teleconferencing take its place as an effective method of communication—not only in the workplace, but beyond.

7

Producing a Satellite Video Conference

by Al Bond

Producing a video conference is a challenging and professionally rewarding assignment. This chapter discusses the mechanics of preparing for a successful satellite conference, what type of conference you can expect to hold with success, and introduces international teleconferencing. It does not examine program content, and the reader is referred to Chapter 6 for a detailed discussion on origination.

OUTSIDE TELECONFERENCING SERVICES VS.
DOING IT YOURSELF

For your first video conference, it is advisable to call on the services of a professional teleconferencing service. You and your staff can learn a great deal the first time by watching them. A professional service knows the correct steps to follow. It has contacts throughout the country (and in some cases throughout the world) for obtaining hardware and transmission at reduced costs, and has experienced staff. Especially if you do not have a video background, your company has never had exposure to internal television, and your management is looking for excellent results from the very first meeting, then working with a teleconference service will provide excellent training for you and will help to ensure a successful video conference.

With all the services available, why consider producing your own conference? You may be limited by budget or personnel, or you may be the type of manager who prefers to do the job himself. Another consideration is that in the current video conferencing market few services are geared to smaller meetings. However, while the multi-thousand dollar conference is today's prize, the small manager-to-manager meeting will affect company profits today and in the future, and I expect that conference services will begin to work on small conferences as well as major ones. At that point the justification to produce your own teleconference will become far more difficult.

Despite its complexity, producing a video conference on your own is not out of the question. However, you must carefully check each step of the production process outlined below to ensure that your meeting does not fail.

A Word About Time

All your planning is affected by time. The more time you have, the more cost-effective your video conference will be. As time is lost to indecision, so are money-saving steps. Leaving six months' lead time for your first meeting is not excessive; on the other hand, leaving only ten days will double your costs because of the fast reaction time required.

Large meetings require careful planning with many suppliers and a well informed staff, and coordination of a great deal of hardware. Long lead time will allow you to check with many vendors to negotiate better prices for most of those items required for the video meeting. Money alone cannot buy services, as even the most experienced suppliers can do a quality job only when given a reasonable amount of time. A responsible manager views video conferencing as a cost-effective means of enhancing the company's productivity. Excessive costs caused by short lead time erase this strategic advantage of video meetings.

IMPORTANT STEPS IN PRODUCING A TELECONFERENCE

The following (see Figure 7.1) is a checklist of "must-do's" to follow as you prepare for a teleconference. Each step will be explained in turn below.

Figure 7.1 Teleconference Production Steps

1. Select origination site
2. Select the right satellite
3. Contact time owner; order time
4. Arrange transmission to uplink
5. Reserve downlink(s)
6. Select viewing location(s)
7. Select viewing hardware
8. Coordinate audio
9. Select on-site coordinators
10. Evaluation

SELECTING THE MEETING SITE

In many cases, selecting the origination site is not an option—often, it has been preselected by management. When management has not specified the origination point and that choice is yours, an existing video facility should be used for a large meeting. An established facility will be properly lighted, acoustically correct and designed to take less set-up time. For a smaller conference it is best to originate from an area close to the normal work environment.

As for any type of conference, it is imperative that you understand who makes up your audience and what it expects. Although all audiences are in some ways similar, it is clear that upper management personnel perceive themselves as requiring different treatment from the average worker. "Spacious," "comfortable" and "well-equipped" are key words for the executive meeting. "Serviceable," "compact" and "accommodating," on the other hand, depict the environment of lower-level management and staff meetings.

CHOOSING THE RIGHT SATELLITE

A communications satellite orbits the earth 22,300 miles above the equator in a geosynchronous (or stationary) orbit. This means that it travels at the same speed at which the earth is orbiting and therefore, relative to an observer on earth, does not move from the point above the equator where it is parked following its launch.

Each satellite has receiving and transmitting antennas, called *transponders*. Your request will be for a transponder on a satellite, not for the entire satellite. Satellite owners sell transponder time in blocks of years for millions of dollars. Some major time buyers, such as Home Box Office, use all their time. Others, like Hughes Television, make satellite time available to the general public and lease it on request. Each satellite has a number of transponders, and the owner will lease you time on any available transponder. For all practical purposes, most satellites will have a similar coverage area, or footprint. Available time is your primary concern.

Major satellite owners, such as Western Union, do have *occasional time* available and will rent it for less than that offered by major purchasers. This, of course, is like buying from the manufacturer rather than the local dealer. Try to buy direct from the owner, but remember that the time available is called occasional time and it may not meet your meeting needs.

What is the right satellite? The right satellite is one that meets the following conditions:

- Time available when you need it;

- Uplink facilities permanently installed within a reasonable proximity of the originating site;

- Coverage to meet your needs.

The wrong satellite is one that has any of the following problems:

- When you can clear only part of the time needed or just the exact time requested, not allowing for an "approximate" end of the meeting. This boxes your meeting into an unsatisfactory constraint.

- If the backup uplink is hundreds of miles away from the point of origin, you will incur high terrestrial charges to move the picture over the earth.

- If a portable uplink is required it will be expensive. You should compare this cost to the cost for longlines transmission to permanent uplinks.

- Some transponders are more difficult to receive in the far corners of the U.S. than others. To be certain that the coverage is adequate in your reception area, check with downlink suppliers for the local results on a specific satellite and transponder.

ORDERING TRANSPONDER TIME

Obtaining transponder time can be likened to renting a hotel room. Rooms are available but the quantity is limited, and volume users, such as conferences, may occupy all or most of the available rooms. Shopping for the right room at the right price and time is required. It is clear that readily accessible transponder time will decrease in the future, even though more satellites are placed into orbit. You can expect heavy committed usage of transponders as the time of day slips into evening. The note below gives you names and phone numbers of companies holding large blocks of transponder time as of March 1982.* A reputable company will be glad to refer you to others if it is unable to meet your requirements; ask for that referral if it becomes necessary.

Transponder time is, at present, inexpensive for the coverage you get

---

*Contact list: Group W Productions (Western Union): L.C. Doerr (1) (800)245-4463; Hughes (Madison Square Garden): Mike Page (212) 563-8900; NETCOM: Bob Paterson or Bill Gillson (1) (800) 423-2085; PBS: Polly Green or Ralph Sheets (202) 488-5084; RCA: Lou Donato (212) 248-2069; Showtime: Ann Korb (212) 708-1693; Western Union: Mike Caffarel (201) 825-5000; Robert Wold Co.: Mike Sterba or Mark Waller (212) 947-4475.

compared to microwave cost. You can expect to pay from $450 to $750 per hour to reach the entire continental United States. This is for time on the satellite; there are other teleconferencing cost factors, as shown in Chapter 5, yet the price is remarkably low. (In fact, management usually finds it hard to believe that so little money can reach so far until the first meeting has been completed. After management's first experience, the task of selling this part of the meeting cost will be easy.)

When ordering time allow for a pretest. There are two types of order times: actual or approximate. *Actual time* is slightly less expensive but will result in the termination of your video and audio feed regardless of program duration. *Approximate time* allows your meeting to run over time without fear of termination. This request should be made when you place your order, to enable the time owner to plan for other customers' usage.

UPLINKS AND DOWNLINKS—PORTABLE VS. PERMANENT

An *uplink* is the transmission antenna located on the earth which sends the television signal up to the satellite, which, in turn, relays the picture back to earth. Satellites can receive signals from more than one uplink at a time, but not on the same frequency.

A *downlink* is the satellite-receiving antenna. It can be tuned (like a TV set) from one transponder to another and moved from one satellite to another. Downlinks vary in size depending on where they are located and how strong the satellite signal is. For example, the signal strength of U.S. satellites is strongest in the center of the country, with weaker signals at the northern and southern borders. The supplier is responsible for selecting the size of the receiving antenna.

Permanent Uplinks

Once you have chosen the uplink facility, the next vital link is to find a transmission path to it from the origination point. When permanent uplink facilities are being used, microwave transmission or cable/microwave relays are generally used. It should be noted that some meeting facilities are hard-wired (i.e., permanently wired) to the uplink. At present these are relatively scarce, but you can expect to see more develop in the future.

Let's assume that your meeting point is not at a hard-wired point and that microwave transmission must be used. Microwave services may be found in major cities throughout the country. They may be satellite-owning companies, local television stations or independent services. Also, satellite suppliers will be glad to refer you to such services, as it is clearly to their benefit to get your program to their uplinks. AT&T is the most established company in the business of microwave transmission—they have been providing this service for years. If you use AT&T's transmission, it will take your signal from the origination point to a central distribution point (or switching point) called a"toll." Transmission from this point to permanent uplinks will probably be hard-wired.

Costs for microwave transmission service are quite variable and can depend on the following factors: line of sight from point of origin to permanent relay or uplink facilities; cost to construct a microwave transmission system, if necessary; and the competition in the immediate area for this service.

Portable Uplinks

Portable uplinks eliminate the need for microwave or cable transmission to a permanent facility. With these units you can transmit directly to the satellite, thereby increasing the number of potential origination locations.

When ordering satellite time you must indicate that you will be using a portable uplink so that the owner can clear the time for your transmission. Unless time is scheduled, permanent uplinks can interfere with reception, since satellite owners may conduct tests which will destroy your signal by transmitting another signal to the same transponder. (In addition, it is against the law to use a satellite transponder without permission of the owner.)

Why not use a portable uplink rather than permanent? The reason is cost. Rental charges for a portable uplink have always been much higher than for permanent installations. In the past, the daily rental charge for a portable uplink has been as high as $10,000. This compares with only $4000 per day for a permanent uplink. In addition, permanent uplinks can be rented by the hour, while portable uplinks must be rented for the day. The rates for portable uplinks have come down, however, and are likely to decrease in the future as more portable uplinks become available.

Today, the average video conference does not require a portable uplink. Normally, it is required for a video conference that originates at an out-of-the-way location, or uses a satellite transponder which has no permanent uplink facilities in the area.

Downlinks

Downlinks or TVROs (television receive-only earth station) are satellite receiving stations, and are multiplying rapidly. Permanent and portable units enable you to view a program at any location—a factor that has contributed to the increased popularity of satellites.

You can rent portable receive dishes from a manufacturer or major supplier; either will arrange for on-site support during the meeting. On-site backup support, when a person is standing by in case of trouble, will cost from $1250 to $1800 per day. A dealer will not provide on-site backup. The importance and duration of your meeting will determine what type of vendor you choose. Most manufacturers can support all your locations; some, however, are limited to local rentals only.

A TVRO is a large antenna and is not selective to only one type of signal. Any electronic signal in the frequency range of your satellite signal will be accepted by your antenna, resulting in interference levels that can render your picture and sound useless. Whenever a site is selected you must conduct a site survey to ensure interference-free reception. You will either have to pay an additional charge for this or your vendor may include it in the price of rental—in any case it should be budgeted for. A company in the business of renting downlinks can advise you on its previous success in the area you have chosen. Even in questionable areas, selective location of your receive dish can block unwanted signals and ensure good quality video. For example, you can "hide" your dish behind buildings or trees and prevent a radio frequency signal from reaching it, while still maintaining reception of the satellite signal.

Dish size and on-site location are generally up to the supplier. (Incidentally, a satellite dish will run on normal household current.) Normally, a supplier does not feed video signals to the meeting area, but provides audio and video at the dish. Your site coordinator is responsible for providing audio and video lines to the meeting area. Some suppliers will provide only audio and video, still others will provide a modulated signal which is converted to standard TV channels for reception. The type of signal being supplied must be decided at the time of your order to enable the proper selection of display equipment.

When selecting a permanent downlink site remember that at present many such facilities are obligated—that is, they have fixed dishes watching one satellite and therefore can receive from only one satellite and one transponder. Independent, or non-obligated, facilities are generally more flexible since they can receive from any satellite in orbit. This makes it possible for you to specify the satellite and transponder you will be using. With a non-obligated network your options remain open, allowing you a more flexible time schedule, more bargaining room for receive location points and most important, more satellites to select from.

SELECTING VIEWER LOCATIONS

Where will participants and viewers take part in this conference? Location at the normal work environment will facilitate participation and is a positive factor in the cost-effectiveness of the video conference. The use of a nearby conference room for the smaller audience brings the participants into familiar surroundings, while the use of a cafeteria or large shipping or receiving area will suffice for larger audiences. Whenever possible, conferences should be held where workers are located. Bell's Picturephone® Meeting Service has recognized this fact after spending several years trying to bring audiences to its locations. The Bell concept is the most correct for the future, as the smaller meeting will be easier to hold and therefore, more effective. Also, attendance usually increases as you move the originating site closer to participants.

SITE REQUIREMENTS

An audience of less than 20 can easily view a single 25-inch TV. A larger audience must have a large screen television projection system or multiple 25-inch TV sets. In all cases room lighting must be controllable. With large screen projection systems light levels become even more critical. Video projection systems, poor by most standards, produce maximum picture quality at or near the center of a pie-shaped viewing area, with the screen being the point of the pie. The farther left or right of center you look, the poorer the picture quality gets.

Seating should be no closer than 10 feet to a projection system expanding out within a triangle shape. For larger audiences a permanent room-audio system should be used for maximum sound. Special attention must be paid when Q/A (question and answer) is to take place to prevent feedback and reamplification of the same sounds. Microphones must be shielded from speaker systems.

AUDIO COORDINATION

When you buy satellite time you will get audio with your picture. In addition you will need audio for coordination—or simply, the installation of a telephone circuit. You must have a telephone available for coordination at each conference receive point, which can be accomplished by contacting your local telephone representative at each site.

Insist that a business line be installed at your meeting facility; do not rely on a phone that is part of a hotel or conference center system. Direct telephone service from the meeting area to your coordinating area is a must, since it will enable you to respond rapidly when necessary. This phone line will also give the site coordinator more confidence, since he or she will know you are close at hand. This system may double as a part of the two-way audio/one-way video conference when the meeting is underway. You can expect to pay for one month's service even though you may use it for only one day.

Setting up this part of the conference requires the greatest amount of time. Initiating telephone service can easily take a month because of the established request procedures and work requests of the phone company, and the number of companies involved.

If many callers are expected it may be desirable to install a toll-free service with numerous lines. When this type of installation is used, it is essential to have a person on each line in direct contact with the producer who, in turn, has direct contact with the moderator by audio earpiece. This will give the appearance of smooth control by the moderator since he receives directions via the earpiece. Remember, an established voice contact with the site coordinator is a must—early planning will make the conference much more enjoyable.

SELECTING THE SITE COORDINATOR

When you choose a site coordinator you are selecting a vital link in your video conference. Site coordinators are generally area representatives of your company and know the organization very well. Who they are and how they are selected will vary from company to company. In some companies they are salesmen, in others they are facilities coordinators. The site coordinator should have a vested interest in the success of the meeting, and the importance of his or her role should be made clear.

The site coordinator can make or destroy all your planning. Always proceed as though the site coordinator has little understanding of how the conference will take place. The coordinator will not be offended if you prepare a graphic plan giving step-by-step instructions on how to support his or her part of the conference. Explanation of room size, type and style of projection or television receiver, seating, lighting, sound and telephone coordination will clear anxiety in most cases. A drawing of the room layout can be helpful.

A coordinator should always know exactly what is expected of him, and this can best be done by preparing a coordinator's notebook. Items in the coordinator's notebook should include:

- Names of hardware suppliers and when they are expected at the meeting site;

- The satellite and transponder you are using;

- Phone numbers the coordinator may need (usually the suppliers' and yours will suffice);

- A diagram of a typical room set-up;

- An explanation of how the television signal is getting to the meeting site (the audience is always interested in this new technology);

- A pre-meeting agenda of the events that will take place during the meeting; and

- A list of things to say or do if there is a failure of picture or sound.

EVALUATION

Evaluation is a part of any conference, and especially so for teleconferencing, since it is a new medium. Each evaluator will be looking for answers to the same basic questions. Some of these are:

- Did you gain the information you needed?
- Did you like this type of meeting?
- What did you like best?
- What did you dislike about the meeting?
- How would you rate the meeting location?
- How would you rate the equipment used, i.e., television projectors and sound systems?

The best evaluation will come from specific questions that are presented clearly and simply to your audience. Complex questions are hard to understand and harder to evaluate. The good way to get responsible answers is to tie the evaluation to upper management requests and to make it clear that a report will follow reporting your area's evaluation.

TWO-WAY VIDEO CONFERENCING

Producing your first one-way video, two-way audio teleconference will give you the base to expand and explore the video conferencing field. Let me give you a quick feel for the magnitude of a two-way video/two-way audio conference.

Every step mentioned in this chapter must be done twice. For example, you will require two transponders, two receive dishes, two originating sets of video cameras and switchers, multiple television display hardware, dual microwave to uplinks or dual uplinks, and the list goes on.

Clearly, you will be required to run two video conferences at the same time. You should never try to handle a two-way video/two-way audio conference for a "first-time" video conference: the chance of failure is almost inevitable.

It is clear that two-way video/two-way audio conferences will proliferate in the near future. Voice activated audio meeting equipment for two-way meetings is presently available and can be considered if full-time services are to be supported with small audiences of eight to 10 people at each meeting site. However, because of the complexity of the equipment, it is unlikely that this type of hardware will be available for rental soon.

Along with two-way video conferencing, another form of conferencing has gained more attention in recent years: international video conferencing. The following section examines the developments in international conferencing and the roadblocks to its use, and presents a case study of a successful international video conference.

INTERNATIONAL VIDEO CONFERENCING

International video conferencing is an area little explored by the business world, but traveled for some years by the traditional trail blazer, the broadcaster. Broadcasters had a major reason to develop international video: the need to reach to the far ends of the world with picture and sound for news and special events. Once the technology developed, hardware costs dropped and the possibility of international satellite usage was considered for other purposes: most recently, for teleconferencing.

In order to better understand the possibilities of international satellite conferencing it is helpful to review the development of this new and exciting dimension to communications.

International Satellite Communications

In 1962 Congress's Communications Satellite Act was signed into law, mandating the creation of the Communication Satellite Corporation or COMSAT. COMSAT is a corporation owned by shareholders to provide domestic, international and maritime satellite service. In 1974 the International Television Satellite Organization was formed, now known as Intelsat. Originally, 11 countries began this joint venture which has now grown to more than 100 active members. Thus, COMSAT is the U.S. representative in the Intelsat system, and is responsible for all dealings in the U.S.

The Intelsat network has more than 260 earth stations around the world, operating on some 20 different satellites. These satellites carry far more than video programs—voice communications and computer data consume more of the load on today's satellites than do video programs. The impact of video is developing and will come of age as digital compression of video signals allows for more effective use of transponders.

While it may seem that satellite communication is new, in fact it has been with us for nearly 20 years. The first operational satellite was Intelsat I, better known as Early Bird back in 1965. It was used to send the first television transmissions via satellite. Today, satellite technology makes it possible to send video signals from your originating point to almost any location in the world, and at a surprisingly low cost.

Technological Considerations and Roadblocks

As we explore the use of satellites for international video conferencing, new factors must be considered. These are:

- Television standard used at origination point and reception locations;

- Country's ability to transmit signals from receiving point to point of presentation;

- Possible government restrictions on the type of material to be presented; and

- Availability of necessary hardware.

The world is not uniformly the same when it comes to television standards. There are three major standards for television transmission in the world, one of these with significant variations.

In the U.S. the National Television Systems Committee (NTSC) system is in use. Here the signal is based on a format generated with 525 lines horizontal resolution and 60 cycles AC power rate. Throughout most other countries the Phase Alternate Line (PAL) system is used. This system is based on the rate of 625 lines resolution on 50 cycles AC power. It is within this system that we find PAL M which is used in Brazil and is not compatible with other systems of PAL in the world. The third system in popular use is Sequential Colour and Memory (SECAM) which is 625 lines on 50 cycles of AC power.

None of these systems are compatible, which is a major problem. The compatibility problem is solved by the use of translators known as scan converters. However, these are extremely expensive and therefore in short supply.

Once the signal has been scan converted it may be transmitted over native country video carriers. This can present insurmountable problems if the country receiving these transmissions is not equipped to move video as easily as we do in the U.S.

A logical solution to these problems would be to use satellite dishes at the receive point locations. However, in many countries it is not lawful to use a satellite receive dish. For example, in Canada it is currently illegal to watch programming from a U.S. satellite. It is also against the law to program a Canadian satellite for an audience in the U.S.

International Teleconferencing Today

Despite these roadblocks, the technology for international teleconferencing is in place in North America and in many other countries. At least one permanent international video conferencing service is scheduled to begin in early 1983, a joint venture between COMSAT Corp. and Inter-Continental Hotels. Other companies, such as VISCOM and Western Union, offer such services on an *ad hoc* basis (see the Appendix at the end of this book). Following is a case study of one international teleconference.

CASE STUDY: A SUCCESSFUL INTERNATIONAL MEETING

For its 50th anniversary stockholders' meeting, Texas Instruments (TI) decided to bring together more than 9000 employees from such diverse locations as West Texas, Tennessee and Germany. The TI media center staff was assigned the task of organizing this meeting in the most economically feasible manner. After carefully analyzing several alternatives, the staff decided to hold an international teleconference, using two satellites, 11 earth stations, numerous microwave hops and longlines transmission.

The unique feature of this meeting was not the technology. Although some of the capabilities used were new, telephone longlines transmission had been used for the last four years to carry TI's stockholders' meetings. The real surprise was that TI was able to increase the number of remote locations, including Europe, with only a moderate increase in the total cost. This approach was especially attractive to management, since TI is already heavily involved in the manufacture and development of sophisticated communications technology.

Background

TI opens its stockholders' meetings to as many qualified employees as possible to accomplish several objectives. First, it enables upper management to communicate directly with employees. Second, the live nature of the televised meeting helps increase the feeling of active participation, which is especially important at TI since most employees are also stockholders. The meeting is also videotaped for distribution throughout the company. While some of the excitement of the live event is lost on tape, nevertheless extending the meeting to as many locations as possible has increased commitment and involvement with management's goals.

In the case of the 1980 50th anniversary meeting, TI had the additional objective of giving a greater number of employees a chance to participate.

Network Components

After studying the alternatives, and after choosing the additional locations to be added to the network, a plan was developed using a combination of AT&T longlines, a Western Union uplink and COMSAT satellites. The decision to use satellites was not an easy one. In the past, the media center staff had been very successful using Telco, while satellites presented unknown variables. However, satellite transmission made additional sites more accessible. For example, the TI plant located in Johnson City, TN, was physically unable to receive past programs because of the lack of available AT&T hardware. The TI plant outside Houston, TX, had a different problem: microwave costs were prohibitively expensive because of obstructions in the area. By using a satellite transmission both locations could be reached.

In addition, the staff was able to reduce the total cost of the network by selectively using earth stations for certain locations, which represented significant savings. Organizational Media Systems (Dallas, TX) prepared a cost analysis comparing the different alternatives for each site so that costs could be evaluated as the network was put together. This type of information was essential for cost-justifying decisions. A final advantage to using satellites was the capability of reaching Europe. Thus, the reduction in existing network cost as a result of using domestic satellites made TI's first international meeting possible.

Transmission

The plan for transmitting to remote locations was relatively simple, on paper. The meeting originated in a cafeteria at TI's home office in Dallas. The signal was transmitted to other locations in the complex by TI's cable system, which was already in place. Off-site locations were reached through a network, established by AT&T, whose distribution point was at the Dallas toll office. Western Union has an existing uplink in Dallas which made significant savings possible, since microwave transmission was used to the uplink from AT&T toll, as well as to other locations in the Dallas area. If the permanent uplink had been unavailable or nonexistent, the costs would have been much higher. Had the uplink not been available, TI would have been faced with the decision of sending microwave transmission to another city with an uplink in place, or renting a portable uplink for the event—in either case, at considerable expense.

Another area of savings was the rental of earth stations. A number of receive-only stations (TVROs) were rented from Ft. Worth Tower for this one meeting. As a result, TI was able to transmit directly to plant sites without the additional expense of microwave hops from existing facilities. The TVROs were installed with a minimum of problems. They were set up anywhere from two weeks in advance, to the afternoon before the meeting, depending on Ft. Worth Tower's prior commitments and delivery schedule. Picture and audio quality were reported as superior to excellent at all locations. Renting units also allowed TI to experiment with its needs before making any decision about purchase.

Using the satellite and earth stations eliminated one problem that had plagued TI every year—trying to coordinate the hand-off between two separate phone systems. Transmitting from the satellite directly to the plant eliminated this need for coordination between the systems.

Transmitting the signal overseas presented a new set of variables. First, because of government regulations, direct transmission from the Westar satellite to the European transmission satellite was not allowed. In order to reach Europe this signal had to be transmitted from Westar to a city nearest a gateway city—in this case Washington, DC. From Washington, the signals were transferred from Western Union to AT&T, which carried them to Etam, VA (the gateway city) for connection to the COMSAT uplink. The signal was received in Europe and transmitted by microwave to TI plants outside of London and Munich. The other major factor in teleconferencing to Europe is the need for both a scan converter and a receive station accessible to the meeting location. These were in place at the English and German plant sites which made the program concept feasible. The plant near Munich was readily accessible to a downlink and scan converter which had already been installed for the Olympic Games. The plant near London presented no problem since scan conversion and a receiver station were also readily available. This was not the case in other European countries. As a result, TI elected to limit its live transmission to these two locations, where costs could be kept to a minimum.

Planning

The decision to use a combination of satellites and longlines transmission was made about 90 days before the stockholders meeting, which was cutting it very close. Any less time and it would not have been possible to adequately coordinate the logistics for the event. The fact that the media center staff had experience with four previous meetings, and had already initiated site surveys, located earth stations, requested bids with the phone company and tentatively reserved satellite time, allowed them to move as quickly as they did. In general, anyone planning a similar type of transmission to several points should start thorough planning at least six months before the event.

The second element that needs attention early is the identification of cost centers. The major cost centers are:

- Transmission from point of origination to uplink;
- Transmission from uplink to satellite;
- Satellite transmission to TVROs;
- Transmission from TVROs to meeting site;
- Transfer to second satellite for international transmission;
- Equipment rental;
- Remote television facility.

The most expensive links in the chain tend to be getting the signal from the origination point to the uplink, or from the receive station to the remote meeting site. The ideal circumstance would be to hold the remote meetings in a facility with permanent downlink or microwave capabilities, such as a hotel network or AT&T Picturephone® Meeting Service facility. In this case, it was not practical because of the large number of people at each site and the transportation costs involved.

With all of the technology to worry about, it is easy to overlook a third element that needs early attention: local logistics, or on-site coordinators. These people are essential in handling arrangements and making sure that the remote meeting runs smoothly. At TI, plant contacts are selected by facilities representatives at each location to ensure a single source for coordination.

When TI first started televising its stockholders' meeting, almost every location experienced some problems. These problems have been reduced during the past five years, since the media center staff developed a system of performance checks prior to the meeting. These checks eliminate as many problems as possible before the actual telecast. All equipment is checked by the engineering staff two weeks before the meeting. This includes a careful inspection of the control room, four cameras, transmission network and audio systems. A network pre-check is run the evening before the stockholders' meeting to make sure that each location is receiving satisfactory audio and video signals. In the case of its international teleconference, this was especially important since this was TI's first experience with using satellites.

Evaluation

For its first international teleconference, the media center staff did have some coordination problems; however the meeting was not subject to a loss in picture or any other serious breakdowns. The staff now feels much more confident about using satellites, primarily because so many variables are eliminated. Microwave transmission requires a number of hops—the greater the distance, the greater the number of hops—and as a result there are more chances for something to go wrong. Satellite transmission, which is more distance-independent, reduces the number of hops involved—transmission goes from an uplink, to the satellite, back down to a receive station. However, before trying to put together an extensive international network using a variety of transmissions, it is first necessary to have experience in coordinating all aspects of a domestic video conference.

8

Teleconferencing Applications

by Robert D. Rathbun

Since the mid-1970s, advancements in satellite technology, digital signal compression, microwave transmission and various audio and video capabilities, together with the soaring cost of travel and the business community's increasing need for effective communications have all led to the development of highly effective and increasingly accessible methods of teleconferencing.

Currently, the two most common forms of teleconferencing are the audio conference and the video conference. The former is the oldest and simplest method—apart from in-person meetings—for bringing two or three or more persons in different locations together for the purpose of communicating. Audio conferencing is used to transmit information by sound alone. Video conferencing is a relatively new twist to this telecommunications process that requires a high degree of technical sophistication. A video conference, while also capable of transmitting sound, has the additional and distinct advantage of providing visual material as well.

Any organization investigating the possibilities of incorporating teleconferencing into its operations has four options to consider: audio conferencing; enhanced audio conferencing, using slow-scan or compressed video; full-motion one-way video/two-way audio conferencing and full-motion two-way video/two-way audio conferencing. (Chapter 2 describes the technology of each of these formats in detail.) Each format has its advantages and each can be used for a wide range of applications. It is up to the user of the medium to determine which form of teleconferencing will work best for a particular need and whether the cost can be justified by the end result (see Chapter 5, "The Economics of Teleconferencing").

Fast developing technology and tough economic conditions are dictating a wide range of new uses for teleconferencing, and a wide range of new users as well. Applications that a few years ago were only feasible for very large corporations are now feasible for medium and even small organizations as well.

HOW TELECONFERENCING APPLICATIONS DIFFER

Audio Conferencing Applications

Audio conferencing via telephone lines is undoubtedly the easiest and least expensive form of teleconferencing now available. Wherever a telephone is located there is a potential teleconference site. With the growing demand for this kind of service, companies such as The Darome Connection and Connex International have been formed to provide audio conferencing networking capabilities (bridging, etc.). Audio conferences are most effective when the event does not require any visual material, and when it needs to be coordinated on short notice.

Some characteristics associated with the audio conference, however, make it an unacceptable format in many situations. For example, any meeting that would suffer without graphic enhancement would not be effective with this method. Audio conferencing is least effective when face-to-face contact is an important part of the communications process, such as during a sales presentation.

Enhanced Audio Conferencing Applications

Enhanced audio conferencing, incorporating slow-scan or compressed video, is another of the less expensive forms of teleconferencing. It not only offers transmission of audio information but also enables users to send pictures along with the sound. Video transmission via the slow-scan process is limited, primarily because the technology involved (compression of the video signal to a narrow bandwidth) does not have the capacity for handling full-motion pictures. While recent developments in scan-converter technology allow pictures of relatively high resolution to be transmitted, these pictures are still static and slow-scan video is best used for sending graphics and other visual materials that do not require movement.

Nevertheless, the slow-scan process gives participants the ability to make use of limited visual material. Furthermore, even though it only transmits still pictures, slow-scan is significantly less expensive than producing a full-motion video conference. Excluding the cost of equipment, slow-scan conferences are only as expensive as long distance phone rates. Slow-scan is best used when visuals are needed only to support the audio, and are of a stationary nature. When full motion is required—such as when the movement of a piece of equipment is to be shown—slow-scan will prove ineffective.

One-way Video/Two-way Audio Conferencing Applications

Full-motion one-way video/two-way audio is the most widely used of the video conferencing formats for several reasons. First, it is less costly and easier to produce than a two-way video event. Second, it has proved to be an efficient and easy way to bring large numbers of people together for a single presentation. This format can accommodate full-motion video and color pictures of the highest quality, along with complete audio interactivity in any location where satellite downlink capabilities and telephone lines are available.

This style of teleconferencing seems to work best 1) in situations where demonstrations are required and 2) when all the information is coming from a single source. The most common applications for one-way video/two-way audio conferencing are for sales and training meetings, new product introductions, video press conferences and annual meetings, because these events often place heavy importance on visual aids and graphic material.

However, when participants in more than one location must present complicated visual material, this format will not work. Also, the costs involved in setting up a multi-location, full-motion video conference generally require a large number of participants to justify the expense. For small gatherings—even when spread over great distances—it may be preferable to opt for an audio or enhanced audio format because of the cost savings involved. In such cases, the money saved may outweigh the effectiveness of the event.

Two-Way Video/Two-Way Audio Conferencing

Full-motion two-way video/two-way audio is the most versatile of all the teleconferencing formats and always the most difficult to produce. Except in cases where a permanent teleconferencing network with fully equipped transmit/receive locations has been established, such as AT&T's Picturephone® Meeting Service or the Boeing Aerospace Company's in-house microwave network (discussed later in this chapter), this form of teleconferencing can be very costly and rather complicated.

Relying on satellite, microwave or cable transmission of audio and video signals, two-way video/two-way audio conferences can be held between two or more locations and provide the most complete form of interaction short of a face-to-face meeting. When satellite transmission is used teleconferences are extremely expensive and require tight coordination because of the demand for both transmit and receive equipment at each site.

Until further technological advancements bring down transmission costs (e.g., by making more satellites available, at lower cost), this form of teleconferencing will be best for, and found in, only very specialized situations. These are meetings in which it is crucial that all participants be able to see and hear what is being said and shown at each location; in other words, when total information exchange is of the greatest necessity. For example, if groups of engineers or computer programmers are teleconferencing to discuss technical problems that must be corrected, only two-way video/two-way audio conferencing will prove adequate. Thus, this format has, to date, generally been used in unique situations which almost always involve high technology corporations with generous communications budgets (see Chapter 8, which discusses the status of teleconferencing activities).

Although two-way video/two-way audio conferencing is considered the "state-of-the-art" not all teleconference users aspire to it. It is not a simple format to produce—either for a permanent network or on a one-time basis—and it is not always required. Regular teleconference users who need a method for replacing face-to-face meetings that involve textual and graphic material, should consider this format. For others it is a waste of time, money and effort.

The preceeding discussion has dealt with general applications for the various forms of teleconferencing now available. However, the medium's many capabilities are best illustrated by the practical situations to which teleconferencing has been applied. The following case studies demonstrate how various organizations have used teleconferencing to improve communications.

CASE STUDIES IN TELECONFERENCING

Audio Conferencing

Merck, Sharp & Dohme: Telephone Sales Conference

A teleconference is not always an event that is planned months in advance. Sometimes the medium has to be used to help overcome unforeseen situations that threaten the ability of groups to get together for important meetings. One such event was a telephone teleconference quickly carried out by the pharmaceutical company Merck, Sharp & Dohme when an unexpected snowstorm made it impossible for a group of employees to travel to a scheduled sales meeting. Utilizing the telephone conferencing service provided by The Darome Connection (Skokie, IL) the conference leader from Merck, Sharp & Dohme was able to gather all participants via a telephone teleconference in less than one hour. In this case the group leader dialed up all attendees through the Darome service. (However, the service also allows participants to call into one main number, when pre-scheduling time permits, and be hooked into the audio net.)

Guardian Life Insurance Co.

Another example of the rapidity with which audio conferencing can bring groups of people together was a session, also arranged by Darome, for the Guardian Life Insurance Co. The purpose of this event was to disseminate information to individuals at 29 field offices regarding changes in the coverage offered to 20,000 policy holders. First, a six-page report was sent to sales managers at the field offices explaining the particular policy changes that would go into effect. Next, an audio conference was scheduled. On the day of the conference, managers at the 29 offices dialed a special number, set up for networking this particular conference, and in minutes were meeting as a group along with 75 members of the company's national sales team. At locations where there was more than one person taking part, special group conferencing convener units were used to amplify the audio so everyone could hear what was being said. These convener units were also equipped with microphones for two-way communications.

The event cost the insurance company a tenth of what it would have cost just to bring the 29 field managers to the home office in New York for a one-day conference, not to mention what the bill would have been if the 75 members of the national sales team had attended in person.

Enhanced Audio Conferencing

University of Wisconsin-Extension: Slow-Scan Instructional Network

The University of Wisconsin-Extension (UWE) has taught college courses for nearly 20 years via telephone transmission of audio lectures. Recently, the University established a statewide narrowband video instructional network to send slow-scan television pictures along with audio via phone lines to students gathered in 25 different locations. Wisconsin students involved with the audio lecture program now have the benefit of seeing the instructors who are conducting their courses, in addition to receiving visual information in the form of charts, graphs and pictures.

Conventional TV cameras and monitors can be used in conjunction with specially developed scan converters to send visual information. Each of these converters contains two digital picture memories, which enables one picture to be transmitted and held in memory in preparation for showing, when another is already being viewed by the participants. The instructor can then show the second picture, and even switch back and forth between the two. And as the lecture progresses, additional pictures can be sent out over the net and held in the converter's memory until they are shown.

An additional advantage of the system used by UWE is that faculty members can present lectures from wherever they have access to a telephone. A slow-scan TV system, consisting of camera, scan converter, transceiver and telephone modem hooked to a dataphone, is small enough to be shipped to any location for transmitting video signals. A second telephone with a voice amplifier and loudspeaker is used to transmit the audio.

One-way Video/Two-way Audio Conferencing

Ford Motor Co.: New Product Introduction

The Ford Motor Co. has been teleconferencing off and on since 1980 to communicate interactively and simultaneously with its entire dealer network and salesforce throughout the United States and Canada. One of these events was a three-phase video conference used to introduce new Ford vehicles to three different audiences: dealers, salesmen and members of the press.

Originating from the studios of Producers Color Service (Detroit, MI), the teleconference was beamed to 38 cities and lasted a total of four hours. Coordination of the conference was handled by Spotlight Presents (New York, NY), with VideoStar (Atlanta, GA) supplying satellite transmission services on the Comstar D-2 and Westar I satellites. (Spotlight Presents specializes in the coordination of video conferences while VideoStar operates a teleconference meeting service that links hotel and other receive locations in specialized configurations that meet the needs of individual clients.)

The Ford teleconference was divided into three different sessions and included the use of both live and pre-taped material presented to viewers gathered in the Detroit studio and at the receive sites (various hotels throughout the country). The first session, developed for dealers, ran approximately 90 minutes with a viewing audience of more than 5000. Session two, which took place immediately after, also ran 90 minutes and was seen by approximately 15,000 salesmen. Both of these sessions were, in turn, divided into three segments and were presented in a news program/talk show format. A backdrop of color monitors and a chroma key were used to display taped material, produced specially for each viewing group, about the new products. The third segment was produced to give reporters and editors gathered in 21 cities an opportunity to see, hear and ask questions about the new automobile introductions. In all three teleconference segments, two-way audio via phone lines and satellite transmission was provided with a telephone bridge set up at the studio coordinating incoming calls. Two satellites (Westar I and Comstar D-2) were used to ensure against transmission failure if a problem developed with one of the transponders. Additionally, it was easier for some of the receive sites to get a clear signal from one satellite than the other. Other backup precautions included stationing technical personnel at each receive site, a satellite engineer at the origination site, and pretesting of all downlinks and playback equipment.

Maritz Communications, a Detroit-based production company that produces training programs for Ford's video disc player network, handled all on-site video production for this teleconference. Using a five-camera setup at Producers Color Service studio, full-color video signals were sent by Maritz to an uplink provided by Greater Starlink in Detroit and up to the satellites. Each program followed a rehearsed script that was tailored for the particular audience in attendance. All switching and direction was done from the Producers Color Service control room.

To gauge the effectiveness of this event—Ford's fourth video conference that year—Spotlight Presents also developed a post-conference questionnaire to tabulate viewer response to the programs. Overall, the reactions to the event were favorable as the majority of respondents gave the conference a "very good" to "excellent" rating.

West Virginia Rehabilitation and Training Center:
Training Seminars Via Satellite

The West Virginia Rehabilitation and Training Center (WVRTC) started conducting vocational rehabilitation training seminars via satellite transmitted video conferences in February 1982. At that time the organization conducted its first teleconference—a 90-minute program linking 130 participants located among 16 receive sites in the Eastern, Central and Pacific time zones. The event comprised one part of a five-hour training session designed to instruct rehabilitation professionals in competency assessment and development of skill objectives for disabled persons in the areas of employment and personal development.

Ford Motor Co. held a one-way video/two-way audio conference to introduce new products to three different audiences. Courtesy Producers Color Service.

The teleconference opened with a live video feed from a studio in Washington, DC. Information was presented to viewers in both live and pre-taped formats that corresponded with printed material that had been distributed to participants in each location. Local "facilitators" were on hand at the receive sites to ensure that all technical requirements were met for proper video and audio reception, in addition to trainers who were there to conduct follow-up, on-site training sessions when the teleconference was completed.

Under the direction of David Molinaro, training associate for WVRTC, initial preparation for this event began nearly six months prior to the February air date. Along with finding receive sites at locations convenient to the greatest number of participants, preproduction work included the scheduling of time on an available satellite transponder; contacting local facilitators to handle the acquisition and maintenance of technical and support material at each site; the production of a videotaped training program that would be played during the teleconference; coordination of a telephone bridge to accommodate question calls from attendees at the receive locations; and the acquisition of transmission and uplink capabilities.

With the assistance of the Public Service Satellite Consortium (PSSC) (Washington, DC) Molinaro was able to book time on the Westar I satellite. In all but a few cases, public television stations with satellite receive equipment were used to provide viewing facilities. At the other locations, receive sites were provided at schools, hospitals and hotels with earth stations in place.

Because WVRTC could not find adequate studio and uplink facilities in West Virginia, the Bell & Howell Satellite Network studio in Washington, DC, was chosen as the origination site. Bell & Howell also provided all necessary equipment and technical services for the full-motion, color video transmission. Receive and playback equipment at the network sites was coordinated locally by conference facilitators.

The centerpiece of the instructional segment of the teleconference was a presentation, videotaped during a two-day period just prior to the teleconference, from Dr. Richard Walls of the WVRTC staff and professor of education psychology at the University of West Virginia. This presentation was videotaped beforehand to ensure that all necessary information would be covered quickly and satisfactorily during the transmission. A live question-and-answer session followed, during which conference attendees could telephone in to the Washington studio to ask questions and discuss the material. All audio—both phoned-in questions and the subsequent answers—was sent back out over the satellite for all conference attendees at each site to hear. Afterwards, a "news" segment was transmitted to update the rehabilitation specialists on the effects that current government policies were having on their profession, and a final address was made by the director of the National Rehabilitation Association.

The entire project (excluding the cost of recording material presented during the transmission via video tape) cost WVRTC approximately $10,000.

Bache Halsey Stuart Shields, Inc.: Video Press Conference

To debut a new financial service it was planning to offer clients, the brokerage firm of Bache Halsey Stuart Shields, Inc. (New York, NY) produced a video press conference to make a simultaneous announcement of the service in 13 cities throughout the United States. This project, the first for the company, was produced by Satellite Networking Associates (SNA), a New York City-based consulting firm that specializes in coordinating satellite teleconferencing services.

The event was scheduled to originate from New York City with transmission to 12 other sites on Western Union's Westar I satellite. SNA lined up a broadcast studio, uplink and downlink sites, and playback locations, and established phone links with each site to take incoming calls from members of the press attending the conference.

Using existing teleconference facilities wherever possible, SNA relied on Public Broadcasting System (PBS) affiliates in New York and eight other cities to provide satellite receive equipment and video projection systems or monitors. In the four cities where no PBS station was available, hotel accommodations were contracted and TVRO (television receive only) dishes were brought in for the broadcast. SNA coordinated the installation of the necessary equipment and hired local coordinators to ensure that proper reception and operating procedures were followed at each location.

The two-hour conference included a live address from the head of corporate public relations for Bache, and a videotaped program outlining the new service that was to be offered. This was followed by a rotating question and answer session. A phone bridge was established at the PBS studio in New York to receive incoming calls from each site. Calls were taken hierarchically (i.e., first all calls from Phoenix, then all calls from Chicago, etc.) rather than selecting questions randomly from all the sites at once. Audio was then sent back over the satellite to enable questions and answers to be heard by all.

Development and coordination of the video press conference took approximately two weeks. During this period all uplink, downlink and playback sites, studio space and satellite time were contracted. In addition, scripting of the presentation and production of the videotaped segment were completed during this time. The total cost for the project, from start to finish, was approximately $40,000.

MGM/United Artists: Video Publicity Tour

A similar teleconference was produced by the Hilton Communications Network for MGM/United Artists (MGM/UA), to publicize the release of the movie/musical "Pennies From Heaven." Rather than sending the stars and other principals of the movie on a multi-city, multi-week publicity tour, a 15-city one-way video/two-way audio hookup was established on the Hilton Network. (The network, a joint venture of the Hilton Hotels Corp. and the Robert Wold Co., provides teleconferencing facilities at selected locations in the Hilton chain.)

This two-hour event included the film's featured actors, along with the director and screenwriter, all making a presentation from Hilton's facilities in New York City and answering questions from members of the press participating at 14 other locations in the United States and Canada. As with the Bache press conference, questions were phoned in to the broadcast site and then transmitted via satellite, along with answers.

United Technologies Corp.: Corporate Information Program

United Technologies Corp. (UT) chose a one-way video/two-way audio conference in order to introduce a new division—the Microelectronics Center in Colorado Springs, CO—to its employees.

The program originated from an in-house production studio at the Microelectronics Center and was beamed to 11 locations across the country where more than 1100 of UT's employees were watching. The event was a two-hour production utilizing both videotaped material and live presentations to debut the center and explain the company's plans for entering the microelectronics field. The program opened with a videotaped segment describing the acquisition of the facility and its positioning within UT's overall organization. It included a 15-minute series of live presentations from various company officials, and a video "tour" of the plant. This last segment, produced earlier on video tape, enabled viewers to get a firsthand look at the inside of the center rather than having to rely solely on verbal descriptions. A question-and-answer session completed the event as participants at all receive sites were encouraged to call in.

Coordination of the teleconference was handled by Services By Satellite, Inc. (SatServ), a wholly-owned subsidiary of PSSC that offers telecommunications consulting services to business and industry. SatServ had been contracted by Reeves Communications (New York, NY), which worked directly with United Technologies to oversee the fulfillment of all technical requirements for the conference.

Video production at the Colorado Springs origination point was handled by Skaggs Telecommunications (Salt Lake City, UT), an independent video production company. Skaggs sent full-motion, full-color video signals via AT&T land lines to an uplink in Denver, and out to the various receive sites—five United Technologies' facilities, five hotels equipped with transportable earth stations and one hotel with permanent receive capabilities—over the Comstar D-2 satellite. Netcom International (Burbank, CA), a company specializing in satellite services, provided the portable TVROs and also lined up the two hours of satellite time. SatServ coordinated the acquisition of video display equipment at each receive site through local A/V equipment suppliers. It also established toll-free telephone numbers and a phone bridge to accommodate questions from each location, which were transmitted to all receive sites as part of the program. Preproduction time on this event took approximately three weeks.

United Airlines and the Air Line Pilots Association (ALPA): Negotiating Via Teleconference

In 1981, teleconferencing was used to help bring about a settlement when the Air Line Pilots Association (ALPA) and United Airlines hit a snag during contract negotiations. A one-way video/two-way audio conference was arranged which enabled members of ALPA and United Airlines to discuss the contract in complete detail, outline the current state of the airline industry, and explain United's financial situation. Although United had produced and distributed a series of video tapes to its pilots explaining the impending contract talks, both association and company leaders agreed that a live video conference would enable both sides to air all questions and complaints, and therefore bring on a quick settlement.

Preproduction of this event (which began five days prior to teleconference) was handled by Chicago Video (Arlington Heights, IL). Because the conference was scheduled to originate from a hotel in Denver, CO, it was necessary to bring in a portable uplink to the transmission site to beam the video signal to the satellite, Western Union's Westar III. Additionally, TVRO dishes had to be set up at nine downlink sites in New York City, Washington, DC, Fort Lauderdale, Cleveland, Los Angeles, San Francisco, Seattle and two locations in Chicago. PSSC provided uplink capabilities while Netcom International provided TVRO equipment at each receive point.

When first proposed, the teleconference was to have originated from a studio in Chicago. However, circumstances dictated that it take place in

Denver; thus, Chicago Video had to handle the event as a remote production. The company drove its 21-foot mobile truck from its headquarters in Chicago to the Denver site to provide the full-color, multi-camera capabilities necessary. In all, three cameras were used to televise the session. All switching was done live from a control room in the mobile unit parked outside the hotel where the conference was taking place.

The two-way audio was transmitted via phone lines. A voice-actuated public address system was used for discussions between the broadcast site and remote locations, while audio signals were digitally delayed for coordination with the video signals coming over the satellite. A backup audio signal was also transmitted by satellite to safeguard against any malfunctions of the primary audio system. The Darome Connection handled all telephone transmission. During the question-and-answer session that followed presentations by United Airlines' management and ALPA spokesmen, questions from participants were taken from each site on a rotating basis.

In total, more than 1000 pilots participated in the three-hour teleconference. Videotaped copies of the session were duplicated and shown to the other 5000 members of ALPA who were not able to participate.

The relatively short time available for coordinating this teleconference drove production costs up. Consequently, the final price of this event was approximately $100,000.

The Chicago Mercantile Exchange: An International News Conference

The Chicago Mercantile Exchange began trading Eurodollar futures on December 9, 1981. To highlight the first day of business, a trans-Atlantic teleconference was produced to announce the event to members of the European Economic Community (EEC) and the European business press. Presented in a news conference format, the production originated from two locations at the Mercantile Exchange—the trading floor and a conference room—and ran for approximately 90 minutes. Nearly 100 persons attended the conference at the broadcast site while more than 80 others participated in the proceedings from a downlink site in London.

Telemation Productions (Glenview, IL) was in charge of network coordination, on-site video production, acquisition of video equipment for the origination site and supplying technical personnel. Downlinking responsibilities and coordination at the playback site were the responsibility of VisNews International (an international teleproduction company based in Europe). The Eurodollar teleconference was telecast live at 9:40 a.m. Central Standard Time (3:40 p.m. in London).

Unlike domestic teleconferences, this event required special interconnec-

Cameraman focuses on the crowded floor of the Chicago Mercantile Exchange during an international video conference held by the Exchange. Courtesy Telemation Productions.

tion to transmit the video and audio from the United States to England. Signals were transmitted from the Mercantile Exchange in downtown Chicago via a cable/microwave relay to a satellite uplink in Wisconsin. From there they were beamed to the Comstar D-2 satellite and downlinked at a site in New Jersey. From New Jersey, the signals were sent by cable to an uplink in West Virginia, and beamed to Intelsat's Atlantic satellite for transmission to England. The downlink was located in a suburb of London and signals were sent, via the British Post Office, to the playback site at the London Press Center. Because the original video signals were sent in the 525-line NTSC format, conversion to the 625-line PAL standard was necessary for viewing. Signal conversion was accomplished by VisNews.

The on-site video production at the Exchange was a four-camera setup with all production hardware, including a mobile control room, supplied by Telemation. The program opened with a brief introduction describing the start of Eurodollar trading—transmission began at the same time trading started on the floor of the Exchange—followed by statements from various representatives of the organization. The remainder of the program was used to take question from conference attendees on both continents. Audio from the London site came in via trans-Atlantic telephone; answers went back over the cable/microwave/satellite transmission network.

Two-way Video/Two-way Audio Conferencing

Boeing Aerospace Co.: A Permanent Microwave Network

Early in 1980 the Boeing Aerospace Co. conducted a study to determine if video conferencing could be used to cut down on the number of man hours lost when engineers travelled between the company's four manufacturing plants in northwest Washington state for regularly scheduled technical meetings. As part of that study an experimental teleconferencing link using microwave transmission was established among the Boeing facilities, and two-way video/two-way audio conferencing took the place of traditional face-to-face meetings.

Boeing intended to test this experimental network for six months, but after 90 days of operation results indicated that teleconferencing was saving the company substantial sums of money on local travel costs and lost man hours. In fact, the experiment revealed that this system was cost-effective even when it was being used as little as 15% of the available time during an average week.

In June of that year, Boeing management made the decision to operate the network on a permanent basis. The system was expanded to accommodate more video conferencing—up to 1800 hours per year—and technical meetings conducted with all participants gathered in the same place were virtually eliminated. The company then began operating a network of 12 independent teleconference rooms, all equipped with a full complement of video hardware for transmitting and receiving video and audio signals to and from any of the sites in the net. The network uses a terrestrial microwave link via Boeing's Computer Service Network services. Full-motion, color TV signals are sent at costs that are extremely economical, especially when compared to transmission charges for time on a commercial satellite transponder.

Each teleconference room contains a color and a monochrome camera, a simplified camera switching device and a variety of monitors, microphones and audio speakers. Each site can maintain approximately $14,000 worth of video and audio equipment, although some rooms have been built to accommodate larger audiences and therefore require more monitors, microphones and, on occasion, an extra camera. Boeing has attempted to keep the system as simple as possible by using off-the-shelf hardware whenever possible. Because the network is designed to be user-oriented, it is crucial that teleconference participants are able to operate all of the equipment, including cameras and switchers, without the aid of video technicians or production personnel.

While all of the teleconference rooms can be interconnected at the same time, the majority of conferences conducted on the Boeing net involve only two locations. These two-site events provide participants with full two-way video/two-way audio interaction without the need for a master switcher to control picture transitions between participating locations. Two-site conferences operate with participants in each room viewing those in the other room via color observation cameras. Textual materials and data are transmitted by monochrome cameras (designed specifically for this purpose) and displayed on small black-and-white monitors built into the work stations where participants are seated. When more than two locations are involved, video signal switching (to coordinate what is to be seen by all participants) is handled from the main Boeing facility in Renton, WA. For all teleconferences, audio is carried by telephone lines through the use of voice-activated microphone mixers. All microphones are shut off when not in use to eliminate the problems of ambient sound distorting the audio transmission.

A minicomputer was recently hooked into the system to help simplify the procedure of setting up and carrying out all video conferences. When network clients plan a session they are required to schedule conference room time by placing an order with the system's computer, also located at the Renton facility. At the scheduled time of the conference, the computer calls up the appropriate circuits on Boeing's microwave system, opens the

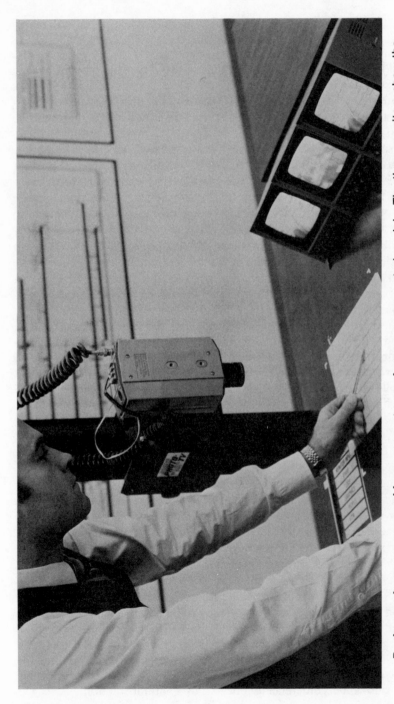

Boeing engineer uses a graphics camera at a conference room control module. The three monitors show the graphics camera and the out-going and incoming signals. A video switcher is near the engineer's left wrist. Courtesy Boeing Aerospace Co.

phone lines and sets the teleconference in motion. It also handles all billing procedures required for internal cost accounting.

A human element has been left in the operation of the net; clients are required to work out scheduling conflicts or problems that may arise when one teleconference runs over into the time slated for another. Although the computer is programmed to know when each session is scheduled to end, it will not shut the circuits down until the conference leader signals that the session is finished.

To spread expenses, the network is operated on a cooperative basis. The annual cost of running the system, and the projected number of hours it will be used, are calculated at the beginning of each year. From these figures the company arrives at the hourly rate it charges its clients. Fees are billed only for rooms that require transmission capabilities, so there is no charge for rooms operated as receive sites only. In mid-1982 the company was billing its in-house clients at the rate of $52 per hour. This low cost can be attributed to the system's use of the already in-place Boeing microwave system for data and voice transmission. Thus, when the teleconference net was first put in place there was no need to construct transceivers at each of the four participating facilities.

Although Boeing's teleconferencing system has worked extremely well, making short distance communications highly cost-effective, it is not a solution for long distance video conferencing. The terrestrial nature of microwave transmission (i.e., transmission is dependent on line of sight and follows the contour of the ground) limits the area that can be covered by such a system. In fact, the longest distance between any two points in the Boeing net is approximately 40 miles.

Aetna Life & Casualty: A Permanent Cable/Satellite Network

Aetna Life & Casualty has created a state-of-the-art turnkey teleconferencing system that presently links its headquarters in Hartford, CT, with a nearby branch office and will soon connect seven other satellite and regional offices throughout Connecticut and across the country.

Late in 1979 the idea for starting an in-house teleconferencing system at Aetna was first raised by computer systems personnel in the company's casualty division. Having recently been moved to a new facility in Windsor, CT—about ten miles from headquarters—they hoped to devise a method of holding regular face-to-face meetings with home office clients without having to make the 20-minute drive between locations. Because of the large number of meetings required each year between the systems analysts and their clients in Hartford, it was apparent that the time and productivity lost during the short drives between the two facilities would be enormous: teleconferencing looked like a natural solution.

By January 1980 the company had conducted studies to develop a video conferencing system that would best meet its needs. A major concern was that the system have the feel of a conference room but provide all the visual and interactive capabilities needed for effective communications.

The mock-up teleconference rooms were built at Aetna headquarters for experimenting with different configurations of video hardware and furniture. Once a design was agreed upon (based on feedback provided by company employees who had practiced with the mock system) plans were drawn up and construction began on the first four teleconference rooms: two in Windsor and two at the home office in Hartford. In March 1981 the system went online and video conferencing began on a regularly scheduled basis.

The company is boasting highly favorable results from its new communications tool. According to Dick Jackson, director of creative services, the first 12 months saw more than 18,000 persons participate in approximately 3000 teleconferences. At an average of 45-minutes per conference, the system is now used about 90% of available time during a given work week, he noted.

All four conference rooms are identical (see illustration, page 179), giving participants on either end of a video meeting the feeling of being close together. Teleconference leaders control camera switching and focusing, and all graphics are displayed via a small control box built into the conference tables.

Each room is equipped with three color cameras—two that are used for general observation and one that can be panned, tilted and zoomed around the room for added coverage. There is also a ceiling mounted black-and-white camera that transmits the graphics displayed on a dry marker tablet built into the conference tables. A 35mm slide projector is available for displaying pictures and other graphics while portable VCRs can be hooked into the system if there is a need for transmitting taped material.

Three monitors are built into the front wall of each room. Two are used for observation of participants and the third is used with the overhead camera, slide chain and VCRs. A facsimile transmission machine is also located in each room for transmitting hard copy. A telephone that can be tied into the system is available for audio conferencing with a third location.

There are five audio speakers in each room—three in the ceiling and two in the front wall—with suppressors that eliminate feedback. Two pressure zone microphones are built into the table so participants can talk without being constrained by wired microphones.

Because the system has been designed for ease of operation, user reaction has been very positive and little technical assistance is generally required in its day to day operations. User orientation is also minimal be-

Participants using one of four Aetna video conference rooms. All four rooms are identical (see illustration on facing page). Courtesy Aetna Life & Casualty Co.

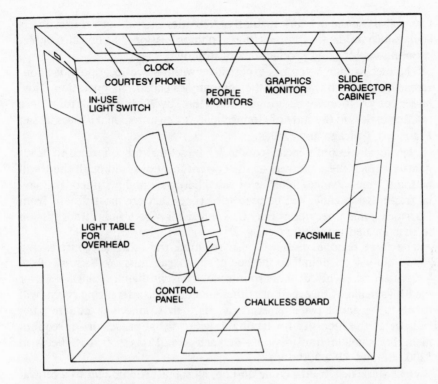

The Aetna video conference room design is depicted in the layout, above. Courtesy Aetna Life & Casualty Co.

cause employees can train each other. The cost savings on travel time alone are estimated at about $400,000 in the first year the system was operating. Since each room cost the company approximately $250,000 to construct, the value of the system is readily apparent. Currently, users are not billed for booking teleconferencing time.

Because the two sites are so close, Aetna has been able to use a coaxial cable transmission system—similar to those used for cable television—to send audio and video signals between rooms. Each room uses 12mhz of the bandwidth available—about three channels—with the audio sent on a sub-carrier. Because it is a closed system, signals are constantly being sent back and forth and no "calling up" special lines or turning on of the system is required each time a new teleconferencing group arrives for a meeting.

Although the net was originally devised for the casualty groups' computer services personnel, it has proved so effective that other divisions in the company have become regular teleconference users. The programmers are the heaviest users but insurance engineers and various management teams also use the network regularly. Activities carried out via teleconferencing include day-to-day meetings, training sessions, technical troubleshooting sessions and product demonstrations.

The success that Aetna has experienced with its two-location video conferencing system has inspired the company to expand its operation. Now under construction are additional rooms that, by 1984, will link four more satellite offices in the state of Connecticut as well as regional offices in San Francisco, Chicago and Dallas.

New local teleconferencing rooms will be set up and connected to headquarters much the same way as the current locations, although there will be some minor changes in terms of video hardware and peripherals. Coaxial transmission cable will be used for offices that are most distant from Hartford while fiber optics may be applied for video/audio transmission to sites located within the city.

The three regional sites will also resemble the other teleconferencing rooms but will differ in their method of signal transmission. Because of the long distances involved, these rooms will transmit digital signals sent over the SBS satellite. Construction of these remote teleconferencing rooms will prove to be about twice as expensive as their Connecticut counterparts because of the necessity for having a digital signal processor—a requirement for satellite transmission—at each site. Processors cost between $200,000 and $250,000 dollars.

Once these new rooms come on-line, all signals will be sent to a central routing switcher located at company headquarters. This will be necessary to control the flow of all signals and to link any and all rooms into whatever conferencing configuration users may require.

CONCLUSION

These examples illustrate that teleconferencing, whether audio or video, has shown itself to be a significant new force in the realm of telecommunications. As its technology continues to advance, the teleconference will become more and more a part of our everyday lives. Not only are major corporations using teleconferencing to communicate, but smaller organizations, non-profit groups and schools and libraries are beginning to see the great potential this medium has to offer.

In the coming years we are sure to see rapid increases in the development of new techniques and applications. The proliferation of smaller, more efficient and more powerful video communications products and the increase in availability of communications satellites, coupled with the economic pressures placed on business and industry to provide greater communications at lower costs, are all factors leading to the acceptance of teleconferencing as a common communications tool, rather than an exercise in futuristic business practices.

But the technology that will emerge for teleconferencing is going to take a back seat to the new and varied teleconferencing applications that will also appear. Today, the teleconference is applied only in situations where it is most obviously needed and/or applicable. As the cost of teleconferencing comes down and the ease with which it can be employed increases, more and more people will find themselves communicating regularly via digital transmission of pictures, sounds and words with associates half a world away.

Conclusions

by Ellen A. Lazer

When Alexander Graham Bell spoke his first sentence into his new invention, the telephone—when V.K. Zworykin of RCA developed the first picture tube—when the first satellite beamed communications from space—no one envisioned the world of teleconferencing as it is growing today. Newer, more sophisticated versions of the telephone, television and satellite communications—as well as numerous other technologies, such as conference bridges, codecs, facsimile devices, remote access slide projectors, etc.—are being developed to make teleconferencing easier and more useful for greater numbers of users.

SOME IMPORTANT CONSIDERATIONS

In order to understand better how and where this medium will succeed—and what the obstacles to its success are—it is helpful to examine some of the factors influencing its development.

Technological advances, not surprisingly, are integral to teleconferencing's development. As the authors of Chapter 2 point out, while much of the technology for teleconferencing is in place, a number of important advances both need and are likely to take place in the next five to 10 years. These developments would, for example, make terminals compatible and portable; integrate audio and video conferencing capabilities with each other and with office automation systems; lead to digital termination systems; and resolve the "last mile" problem. In addition to making teleconferencing more accessible to a greater number of users, such breakthroughs would likely result in a decrease in teleconferencing costs. This would make the medium more attractive to those who presently find the cost of teleconferencing—particularly video conferencing—prohibitive.

While it is primarily developments in technology that are making teleconferencing attractive at this time, technology, we know, is not enough. There are economic and human factors at work here, too. Principal

183

among these is the growing recognition of the need to improve managerial and professional productivity. This need prompts some people to view teleconferencing as a marvelous means of reducing the time spent in traveling to and from face-to-face meetings and perhaps even as a means of leading more effective meetings that result in better decision making. Of easy and sometimes greater appeal is the idea that teleconferencing will significantly redi.ce travel expenditures, a reason frequently promulgated by vendors and consultants of teleconferencing services. Of course, awareness of higher travel, energy and personnel costs, and other economic concerns in a recessionary economy, contribute to this view. However, what portion—if any—of the billions of dollars spent on business-related travel will actually be saved after the direct expenses of teleconferencing are added in, and the potential for holding additional (electronic) meetings is considered, is debatable. Then, too, the possibility of overkill—holding an augmented audio conference when an interoffice print memorandum will do—should not be overlooked.

Not to be underestimated as we think about the future of teleconferencing is the growth of a user population increasingly receptive (no pun intended) to television, computers and other high technology products and services, and becoming more sophisticated and literate in their use. To some degree, this offsets the usual problems of users' becoming comfortable with a new technology. On the other hand, whether it is the 30-year-old manager of training who introduces his company to teleconferencing, or the experienced telecommunications manager who suggests that a permanent network be established, it is top management which makes the decision—and all levels of the company must be educated as to its use.

How soon teleconferencing—particularly two-way video conferencing, the most expensive and most glamorous of these technologies—will create the billion dollar market predicted by some futurists is open to question. While equipment costs have declined over the last decade, transmission costs must decline even more dramatically for video conferencing to become cost-effective for any but a minority of large firms and organizations. It is interesting to note that the recent growth has been in *ad hoc* video conferencing, with fairly flat growth in permanent place video conferencing.

As reported in Chapter 4, among our respondents audio conferencing is the most common form of teleconferencing—more than all the others combined, due to the lower costs associated with audio conferencing, the long-established nature of the telephone network and the relative ease with which an audio conference can be initiated. Continued improvements in audio conferencing technology—in the areas of audio quality and enhancements such as slow-scan and freeze-frame equipment—are likely.

Whether this will prolong user preference for audio conferencing over video conferencing remains to be seen.

THE FUTURE OF TELECONFERENCING

As James Johnson and others have pointed out in this book, it seems certain that teleconferencing will take its place as one of many communications tools rather than being viewed as a substitute for travel or for other means of communication. How soon it will become a routine communications tool is open to debate.

At present, organizations that have implemented video conferencing systems share certain characteristics: they are large, are geographically dispersed and have existing communications networks. Organizations that have held *ad hoc* single event teleconferences cannot be as easily characterized—they vary with respect to size and purpose for teleconferencing. The future may see the video conferencing market break down further along these lines. It also seems safe to suggest that many medium and even small organizations will find it advantageous to engage in some form of teleconferencing at least some of the time.

The U.S. is the acknowledged leader in teleconferencing; within the last decade, demonstrations and tests have generated ongoing applications of this new communications tool. News about the first press conference, "tele-course," medical diagnosis or business meeting held via teleconference is being replaced by news about companies installing permanent teleconferencing networks, hypotheses of what makes a teleconference effective and frequent announcements of new companies (and offshoots of existing companies) reaching to serve the *ad hoc*—or permanent—audio or video conferencer. Organizations, such as the ITCA, are forming to serve teleconferencing users.

Interest in teleconferencing is burgeoning, especially in the U.S. but in other countries as well. Some suggest that as communications technology becomes more sophisticated, as teleconferencing's benefits and applications become better identified, teleconferencing may become as ubiquitous a business tool as the telephone. Others counter that obstacles—such as cost and user resistance—will continue to hinder its widespread acceptance by all but major corporations. This book does not claim to be able to predict what path teleconferencing as a whole will take, nor what form, if any, will prosper over the others. Rather its aim has been to provide a framework for understanding the many aspects of teleconferencing—its technology, applications and implementation—in order that the reader may assess accurately the developments in this dynamic field that are sure to unfold in the future.

Appendix: Directory of Teleconferencing Suppliers

In the summer of 1982 Knowledge Industry Publications, Inc. identified more than 200 companies that were likely to be involved in audio or video conferencing and asked them to provide information about their services. The 113 companies who responded are listed on the following pages in alphabetical order. Following the company's name, address and telephone number are one or more letter codes which represent products or services the company provides.

This directory provides a reference to companies who provide teleconferencing products and services. Of the companies listed, nearly 40 offer full-service video conferencing; more than 30 consult and design video conferencing systems; almost 30 provide creative and production services for video conferencing and more than 20 offer full-service audio/visually augmented conferencing.

We tried to present all relevant teleconferencing services in our Master Key. Almost 30 companies augmented the keyed listings with additional products and services; among them a random access color microfiche projector, slow-scan video products, video hard copy recorders, microwave transmission equipment and an encryption system.

Please note that this information is based solely on self-reporting by each company, and that the teleconferencing business is growing and changing so rapidly that the information in this directory should be used in conjunction with current trade publications and personal research.

As a service to teleconference users, KIPI will in the future issue an updated directory of teleconferencing services. Readers are requested to fill out the form at the end of this section with comments about the directory, suggestions for additional listings, or changes in existing listings.

Master Key

SERVICE OR PRODUCT	CODE
General	
Turnkey (full service) video conferencing	A
Turnkey (full service) audio/visually augmented conferencing	B
Consultant/systems design — video conferences	C
Consultant/systems design — audio conferences	D
Video conference creative/production services	E
Training	F
Transmission	
Satellite time broker	G
Satellite networking services supplier	H
TVRO (earth station) supplier	I
Microwave services supplier	J
Cable interface services supplier	K
Audio bridge supplier	L
Audio bridging services supplier	M
Turnkey network services supplier	N
Common carrier	O
Teleconference rooms	
Video conference rooms (rent)	P
Video conference rooms (build)	Q
Video conference rooms (design/consult)	R
Audio conference rooms (rent)	S
Audio conference rooms (build)	T
Audio conference rooms (design/consult)	U
Terminal (end-user) equipment	
Monitors/cameras/switchers/signal processors (rent)	V
Monitors/cameras/switchers/signal processors (sell)	W
Projectors (rent)	X
Projectors (sell)	Y
Character generators (rent)	Z
Character generators (sell)	AA
Speaker-phone type devices/conference telephones (rent)	BB
Speaker-phone type devices/conference telephones (sell)	CC
Audio equipment (microphones, amps, mixers, etc.) (rent)	DD
Audio equipment (microphones, amps, mixers, etc.) (sell)	EE
Turnkey audio systems (rent)	FF
Turnkey audio systems (sell)	GG
Electronic blackboard	HH
Graphics tablets	II
Facsimile transmission	JJ
Slow-scan (freeze-frame) TV	KK
Newsletters (specify: _____)	LL
Other (specify:_____)	MM

ADDA Corp.
1671 Dell Ave.
Campbell, CA 95008
(408) 379-1500
Services and Products: KK, MM
(sell signal processors)

AKG Acoustics, Inc.
77 Selleck St.
Stamford, CT 06902
(203) 348-2121
Services and Products: EE

All Mobile Video, Inc.
630 Ninth Ave.
New York, NY 10036
(212) 757-8919
Services and Products: V, X, Z,
DD

**Appalachian Community Service
Network (ACSN)**
1200 New Hampshire Ave., NW
Suite 240
Washington, DC 20036
(202) 331-8100
Services and Products: A, E, F,
G, H, K

ATI Telemanagement
120 Tremont St., #323
Boston, MA 02108
(617) 426-0550
Services and Products: C, D

**AT&T Picturephone® Meeting
Service***
Room 3A120
Bedminster, NJ 07921
(201) 234-7879
Services and Products: A, B, C,
D, H, I, J, L, M, N, O, P, R,
S, T, U, BB, CC, DD, EE,
FF, GG, HH, II

Auscom Associates
30-16 Broadway (Rte. 4)
Fair Lawn, NJ 07410
(201) 796-0600
Services and Products: A, C, I,
V, W, X, Y, Z, DD

Austin Satellite Television, Inc.
7320 Mopac Expwy. N.
Suite 404
Austin, TX 78731
(512) 346-2557
Services and Products: A, C, J,
N, P

AVANTEK
481 Cottonwood Dr.
Milpitas, CA 95035
(408) 946-3080
Services and Products: I, J

Avekta
370 State St.
Brooklyn, NY 11217
(212) 852-7568
Services and Products: E, MM

---

*As of January 1983, as a result of a 1980 FCC decision, American Bell, Inc. (the unregulated company formed by AT&T) will be responsible for all equipment, teleconference room rental/building and Picturephone® Meeting Service; AT&T (the regulated company) will retain responsibility for video conference consulting and all aspects of transmission, but will be prohibited from recommending specific equipment vendors.

Avtec Industries, Inc.
39 Industrial Ave.
Teterboro, NJ 07608
(201) 288-6130
Services and Products: C, D, F,
   I, K, Q, R, T, U, W, Y,
   AA, CC, GG, KK

**Broadcast Equipment Rental Co.
   (BERC)**
6352 DeLongpre Ave.
Hollywood, CA 90028
(213) 464-7655
Services and Products: J, V, Z,
   DD

**Burroughs Corp., Imaging
   Systems Division**
Commerce Park
Danbury, CT 06810
(203) 794-9254
Services and Products: JJ

CEAC, Inc.
1500 E. Conecuh St.
Union Springs, AL 36089
(205) 738-2000
Services and Products: BB, CC,
   DD, FF, MM (audio conference bridge)

Channel One
79 Massasoit St.
Waltham, MA 02154
(617) 899-1025
Services and Products: A, B, I

Chicago Video, Inc.
48 E. University Dr.
Arlington Heights, IL 60004
(312) 577-2430
Services and Products: E, V, X

Colorado Video, Inc.
PO Box 928
Boulder, CO 80306
(303) 444-3972
Services and Products: KK

Color Leasing, Inc.
330 Rte. 46 (E)
Fairfield, NJ 07006
(201) 575-1118
Services and Products: A, B, E,
   P, Q, R, V, W, X, Y, Z,
   AA, DD

COMEX
559 Pepper St.
Monroe, CT 06468
(203) 268-5849
Services and Products: MM
   (telephone swing arm)

**Commercial Telecommunications
   Corp. (COMTEL)**
3130 Skyway Dr., Building 604
Santa Maria, CA 93455
(805) 928-2581
Services and Products: C, D, I

**Communications Technology
   Management, Inc.**
6861 Elm St.
McLean, VA 22101
(703) 734-2724
Services and Products: H

**Communication Training
   Consultants, Inc.**
450 Park Ave.
New York, NY 10022
(212) 688-3229
Services and Products: F, MM
   (consultation and coordination)

Compact Video Services, Inc.
2813 W. Alameda Ave.
Burbank, CA 91505
(213) 840-7000
Services and Products: E, G, H,
I, O, P, S, V, X, Z, DD

Compression Labs, Inc.
2305 Bering Dr.
San Jose, CA 95131
(408) 946-3060
Services and Products: V, W

Compucon, Inc.
PO Box 401229
13749 Neutron Rd.
Dallas, TX 75240
(214) 233-4380
Services and Products: C, D, H, J

**COMSAT Corp./Inter-
Continental Hotels**
950 L'Enfant Plaza SW
Washington, DC 20024
(202) 863-6562
Services and Products:
A (international)

**CONFERSAT/Public
Broadcasting Service**
475 L'Enfant Plaza, SW
Washington, DC 20024
(202) 488-5000
Services and Products: A, E, G,
H, I, J, K, N, P, V, X, Z

ConferTech International Inc.
240 Park Center
8795 Ralston Rd.
Arvada, CO 80002
(303) 420-0427, (800) 525-8244
Services and Products: D, F, L,
M, U, BB, CC

Connex International, Inc.
12 West St.
Danbury, CT 06810
(203) 797-9060, (800) 243-9430
Services and Products: B, D, F,
L, M, U, BB, CC, FF, GG

CONTECH
30 Plaza Dr.
Westmont, IL 60559
(312) 789-0888
Services and Products: D, L, U,
BB, CC, GG, LL
(*Contechgram*)

Controlonics Corp.
5 Lyberty Way
Westford, MA 01886
(617) 692-3000
Services and Products: CC, DD,
LL (*Westford Wavelength*)

Cross Information Co.
934 Pearl St., Suite B
Boulder, CO 80303
(303) 499-8888
Services and Products: C, D, N,
R, HH, MM (computer
teleconferencing, software
and services)

DAROME
5725 E. River Rd., Suite 782
Chicago, IL 60631
(312) 399-1610
Services and Products: L, M, U,
BB, CC, DD, LL
(*Connections*)

Editel (joint venture of Bell &
Howell and Columbia Pictures)
985 L'Enfant Plaza, SW
Washington, DC 20024
(202) 484-9270
Services and Products: A

Educational Technology, Inc.
2224 Hewlett Ave.
Merrick, NY 11566
(516) 623-3200
Services and Products: A, B, C,
   D, F, Q, R, T, U, W, GG,
   KK, MM (sells and rents
   teleconference control random
   access slide projectors)

Electronic Systems Products, Inc.
1 Tico Rd.
Titusville, FL 32780
(305) 269-6680
Services and Products: C, F, R, Y

Exxon Office Systems Co.
777 Long Ridge Rd.
Stamford, CT 06902
(203) 329-5000
Services and Products: JJ

Fort Worth Tower Co.
1901 E. Loop 820 S.
PO Box 8597
Fort Worth, TX 76112-0597
(817) 457-3060
Services and Products: A, B, I

Gardiner Communications Corp.
3605 Security St.
Garland, TX 75042
(214) 348-4747
Services and Products: I, K, N,
   AA

General Television Network
13225 Capital Ave.
Oak Park, MI 48237
(313) 548-2500
Services and Products: A, C, E,
   F, P, Q, R, V, W, X, Y,
   Z, AA

Harris-Satellite,
Communications Div.
PO Box 1700
Melbourne, FL 32901
(305) 724-3000
Services and Products: H, I, J, N

**The Hilton Communications
   Network,** in association with
   The Robert Wold Organization
9880 Wilshire Blvd.
Beverly Hills, CA 90210
(213) 278-4321
10880 Wilshire Blvd.
Los Angeles, CA 90024
(213) 474-3500
15 Central Park West, Suite 1601
New York, NY 10023
(212) 247-2120
Services and Products: A, C, E, I,
   J, N, P

HI-NET Communications, Inc.
3796 Lamar Ave.
Memphis, TN 38195
(901) 369-5348
Services and Products: A

Honeywell Inc.,
Test Instruments Div.
4800 E. Dry Creek Rd.
PO Box 5227
Denver, CO 80217
(303) 773-4700
Services and Products: MM
   (video hard copy recorders)

**Hughes Electronic Devices Corp.
   (HEDCO)**
PO Box 1985
Grass Valley, CA 95945
(916) 273-9524
Services and Products: W

Interand Corp.
666 N. Lake Shore Dr.
Chicago, IL 60611
(312) 943-1200
Services and Products: A, B, C,
E, R, W, AA, HH, II, KK,
MM (interactive freeze frame
and graphics pilot programs;
interactive graphics for video
conferencing; freeze frame
networking software package
for multipoint conferencing)

**International Telecom
Systems, Inc.**
122 E. Johnson St.
Madison, WI 53703
(608) 255-8751
Services and Products: B, D, L,
M, U, BB, CC, FF, GG

**J.M.P. Videoconference Group,
Jack Morton Productions, Inc.**
830 Third Ave.
New York, NY 10022
(212) 758-8400
Services and Products: A, B, E, F

JVC Co. of America
Professional Video Division
41 Slater Dr.
Elmwood Park, NJ 07407
(201) 794-3900
Services and Products: W, Y, MM
(remote control unit for
cameras)

Kellogg Corp.
5601 S. Broadway
Littleton, CO 80121
(303) 794-1818
Services and Products: F, M

Keltronics Corp.
136 E. Hill St.
Oklahoma City, OK 73105
(405) 524-2144
Services and Products: L

KPR Infor/Media Corp.
605 Third Ave.
New York, NY 10158
(212) 878-3700
Services and Products: A, B

Larus Corp.
473 Sapena Court, Unit 19
Santa Clara, CA 95050
(408) 727-0583
Services and Products: K, L

LSI Avionic Systems Corp.,
(a subsidiary of Lear Siegler, Inc.)
32 Fairfield Pl.
West Caldwell, NJ 07006
(201) 575-8000
Services and Products: A, C, N,
R, V, X, Y, MM, (large-screen
TV projectors)

M/A-COM DCC, Inc.
11717 Exploration La.
Germantown, MD 20874
(301) 428-5603
Services and Products: G, H,
MM (turnkey satellite networking equipment and service)

M/A-COM Video Systems, Inc.
63 Third Ave.
Burlington, MA 01803
(617) 272-3100
Services and Products: C, D, J,
LL (*Microwave Planning Guide,
Five Minute Guide to FCC
Licensing of Microwave
Systems*), MM (microwave
transmission equipment)

MARCOM
PO Box 66507
Scotts Valley, CA 95066
(408) 438-4273
Services and Products: A, B, C,
  D, E, F, I, J, K, Q, R, T,
  U, W, AA, CC, GG, KK, MM
  (selfcontained mobile
  teleconference systems)

Maritz Communications Co.
4925 Cadieux Rd.
Detroit, MI 48224
(313) 882-9100
Services and Products: E, F

**Maryland Center for Public
  Broadcasting**
11767 Bonita Ave.
Owings Mills, MD 21117
(301) 337-4086, (301) 356-5600
Services and Products: A, E, P,
  LL *(Telecom Update)*

Media Sense, Inc.
1829 Spruce St.
Boulder, CO 80302
(303) 449-0211
Services and Products: C, D, E, F

Microdyne Corp.
PO Box 7213
Ocala, FL 32672
(904) 687-4633
Services and Products: A, C, F, I

Midwest Corp.
Communications Systems Div.
1 Sperti Dr.
Edgewood, KY 41017
(606) 331-8990
Services and Products: C, I, Q,
  R, V, W, X, Y, Z, AA, DD

Misar Industries
17192 Gillette Ave.
Irvine, CA 92714
(714) 540-2477
Services and Products: A, C, O,
  R, MM (audio/video full
  motion equipment)

**Gary Moore Video & Film
  Production**
4908 Arctic Ter.
Rockville, MD 20853
(301) 933-0710
Services and Products: A, C, E,
  P, V, Z, DD, FF, MM
  (mobile production van)

NEC America, Inc.
Broadcast Equipment Div.
130 Martin La.
Elk Grove Village, IL 60007
(312) 640-3792
Services and Products: W, KK

NETCOM International
1702 Union St.
San Francisco, CA 94123
(415) 921-1441
Services and Products: A, C, G,
  H, I, J, N, R, X

Oak Satellite Corp.,
VideoNet Div.
21031 Ventura Blvd., Suite 907
Woodland Hills, CA 91364
(213) 999-3113
Services and Products: A, C, E,
  F, H, I, J, N, R, MM
  (encryption system)

One Pass, Inc.
1 China Basin Building
San Francisco, CA 94107
(415) 777-5777
Services and Products: A, B, E,
F, G, H, J, P, R, S, U,
V, X, Z, BB, DD, HH

Optel Communications, Inc.
(formerly FTC Services)
90 John St.
New York, NY 10038
(212) 669-9700
Services and Products: B, T, U,
FF, GG, II

Panafax Corp.
185 Froehlich Farm Blvd.
Woodbury, NY 11797
(516) 364-1400
Services and Products: JJ

Panasonic Co.
Video Systems Div.
1 Panasonic Way
Secaucus, NJ 07094
(201) 348-7000
Services and Products: R, W, Y

Peirce Phelps, Inc.
2000 N. 59 St.
Philadelphia, PA 19131
(215) 879-7171
Services and Products: A, Q, R,
T, U, W, Y, AA, GG,
II, JJ, KK

Precision Components, Inc.
1110 W. National Ave.
Addison, IL 60101
(312) 543-6400
Services and Products: CC

Producers Color Service, Inc.
24242 Northwestern Hwy.
Southfield, MI 48075
(313) 352-5353
Services and Products: A, P, V,
X, Z, DD, MM (uplink)

Public Broadcasting Service
See: CONFERSAT

**Public Service Satellite
Consortium (PSSC)**
1660 L St., NW, Suite 907
Washington, DC 20036
(202) 331-1154
Services and Products: A, B, C,
D, E, F, G, H, N, MM
(turnkey engineering procure-
ment and construction supplier)

Rauland-Borg Corp.
3535 W. Addison St.
Chicago, IL 60618
(312) 267-1300
Services and Products: D

Revox Systems
(Division of ETI)
2224 Hewlett Ave.
Merrick, NY 11566
(516) 623-3200
Services and Products: Y, MM
(random access color microfiche
projector adapted to
teleconference control)

Robot Research, Inc.
7591 Convoy Ct.
San Diego, CA 92111
(714) 279-9450
Services and Products: KK, MM
(slow-scan video products
mainly for dial-up telephone
network)

ROCKWELL/WESCOM, Inc.
8245 Lemont Rd.
Downers Grove, IL 60515
Services and Products: L

Satellite Business Systems (SBS)
8283 Greensboro Dr.
McLean, VA 22102
(703) 442-5000
Services and Products: A, H

**Satellite Communications
    Network**
Glen Rock Plaza, Suite 103
266 Harristown Rd.
Glen Rock, NJ 07452
(201) 652-0059
Services and Products: I, O

**Satellite Networking
    Associates, Inc. (SNA)**
10 E. 40 St., 23 floor
New York, NY 10016
(212) 889-6460
Services and Products: A, B, C,
    E, G, H, I

**Satellite Transmission
    Systems, Inc.**
80 Oser Ave.
Hauppauge, NY 11787
(516) 231-1919
Services and Products: A, B, H,
    N, MM (satellite ground
    communications equipment)

S.C. ETV Network
2712 Millwood Ave.
Columbia, SC 29250
(803) 758-7261
Services and Products: A, E, J,
    P, S

Scientific-Atlanta
3845 Pleasantdale Rd.
Atlanta, GA 30340
(404) 441-4311
Services and Products: I, J, MM
    (hardware for satellite transmis-
    sion of video or audio signals;
    earth station network design)

Services by Satellite, Inc./Satserve
(a subsidiary of PSSC)
1660 L St., NW, Suite 906
Washington, DC 20036
(202) 331-1960
Services and Products: A, B, C,
    D, E, F, G, H, N, LL
    (*PSSC Newsletter*), MM
    (turnkey engineering procure-
    ment and construction
    supplier)

**Skaggs Telecommunications
    Service**
5181 Amelia Earhart Dr.
PO Box 27477
Salt Lake City, UT 84127
(801) 539-1427
Services and Products: A, E, F,
    V

George R. Snell Associates, Inc.
269 Sheffield St.
Mountainside, NJ 07092
(201) 654-8855
Services and Products: C, D, R,
    U

Sony Corp.
Video Communications Division
Sony Dr.
Park Ridge, NJ 07656
(201) 930-1000
Services and Products: W, DD,
    EE

South Coast Video
(formerly Airborn Media Services)
4500 Campus Dr., Suite 540
Newport Beach, CA 92660
(714) 556-1315
Services and Products: E, P, S,
DD

**Southern Pacific
Communications Co.**
1 Adrian Court
Burlingame, CA 94010
(415) 692-5600
Services and Products: O

**Southern Satellite Systems,
Inc. (SSS)**
PO Box 45684
Tulsa, OK 74145
(918) 481-0881
Services and Products: C, D, G,
H, O, LL (uplinking from
transportable earth station)

Spotlight Presents
20 E. 46 St.
New York, NY 10017
(212) 986-5520
Services and Products: A, B, C,
D, E, F, H, N

Talos Systems, Inc.
7419 E. Helm Dr.
Scottsdale, AZ 85260
(602) 948-6540
Services and Products: B, II, JJ

Telautograph Corp.
8700 Bellanca Ave.
Los Angeles, CA 90045
(213) 641-3690
Services and Products: JJ

Telecasting Services
1058 W. Washington Blvd.
Chicago, IL 60607
(312) 738-1022
Services and Products: B, E, I, P,
X, Y

TeleSession Corp.
475 Fifth Ave.
New York, NY 10017
(212) 889-8066
Services and Products: M, S

Tellabs Inc.
4951 Indiana Ave.
Lisle, IL 60532
(312) 969-8800
Services and Products: L

Uni-Linx Telephone Conferencing
853 Broadway
New York, NY 10003
(212) 674-3588
Services and Products: B, D, L,
M, U

University of Nebraska-Lincoln
PO Box 83111
Lincoln, NE 68501
(402) 472-3611
Services and Products: A, C, E,
H, P, V, X, Z, BB, DD

Vidcom
2426 Townsgate Rd.
Westlake Village, CA 91361
(213) 991-1974
Services and Products: A, B, E,
P, Q, R, V, Z, II, KK

Videomeetings
2636 Walnut Hill La., Suite 110
Dallas, TX 75229
(214) 358-2021
Services and Products: X

Videonet
21031 Ventura Blvd., Suite 505
Woodland Hills, CA 91364
(213) 999-3113
Services and Products: A, C, E,
F, N, P, R, MM (encryption
services)

VideoStar Connections, Inc.
3390 Peachtree Rd.
Atlanta, GA 30326
(404) 262-1555
Services and Products: A, B, C,
G, H, I, J, N

Video Technology Resources, Inc.
1616 Soldiers Field Rd.
Boston, MA 02135
(617) 254-2110
Services and Products: A, C, D,
Q, R, T, U, W, AA, GG, KK

VIDICOM
742-D Hampshire Rd.
Westlake Village, CA 91361
(213) 889-3653
Services and Products: A, B, C,
D, F, L, R, U, W, Y, AA,
BB, CC, FF, GG, II, JJ, KK

VISCOM International
International Bldg.
Rockefeller Center, 630 5th Ave.
New York, NY 10111
(212) 307-7315
Services and Products: A
(international)

Western Union VideoConferencing, Inc.
1 Lake St.
Upper Saddle River, NJ 07458
(201)825-5419
Services and Products: A, F,

WETACOM, Inc.
955 L'Enfant Plaza, Suite 7200
Washington, DC 20024
(703) 998-2700
Services and Products: A, B, C,
E, G, H, I, K, N, V, X, Z,
BB, DD

Hubert Wilke, Inc.,
Communications Facilities
Consultants
260 Madison Ave.
New York, NY 10016
(212) 578-4646
Services and Products: C, D, F, R, U

WNET/NET Telecon
356 W. 58 St.
New York, NY 10019
(212) 560-2067
Services and Products: A, B, E,
H, N, P, S, V, Y, Z, DD

Wold Communications, Inc.
10880 Wilshire Blvd.
Los Angeles, CA 90024
(213) 474-3500
350 Fifth Ave., Suite 8208
New York, NY 10118
(212) 947-4475
Services and Products: G, H, I, J, N, O

Xiphias
233 Wilshire Blvd., Suite 900
Santa Monica, CA 90401
(213) 399-3283
Services and Products: MM (computer graphics systems)

Teleconferencing Directory — Update

1. Is this directory useful to you? For what reasons? _____

_____

2. What categories of products or services would you like to see added to

this directory? _____

_____

3. Was the format of this directory easy to use? _____

What changes would you suggest? _____

_____

4. What companies or organizations would you like to see added to this

directory? _____

_____

5. Any other comments about the directory: _____

_____

_____

Your name, company and address (optional): _____

_____

Thank you for your interest. Please cut this page along the dotted line and send it to: Ellen Lazer, Knowledge Industry Publications, Inc., 701 Westchester Avenue, White Plains, N.Y. 10604.

Glossary of Terms

by Martin C.J. Elton and David Boomstein

Ambient noise: Background noise.

Analog signal: An electromagnetic wave encoded so that its power varies continuously with the power of a signal received from a source (e.g., a source of sound or light).

Audio conference: A *teleconference* employing only voice communication. Used interchangeably with audio teleconference. See also *enhanced audio conference*.

Audiographic conference: See *enhanced audio conference*.

Bandwidth: The difference between the lowest frequency and highest frequency being transmitted in *analog* form. The bandwidth of a signal cannot exceed the bandwidth of the channel on which it is carried.

Bit: A contraction of "binary digit." It is the basic unit by which computer scientists and engineers measure the amount of information. (However, they define information in such a way as to avoid considering content. Really, bits are used to measure the amount of message rather than the amount of information.)

Bridge: A device for interconnecting three or more transmission channels. There are audio bridges and data bridges. Some audio bridges are of the "meet me" variety; participants call in to them at a prearranged time. Others are operator-assisted; an operator calls the participants.

Note: Words that appear in italics within definitions are defined elsewhere in this glossary.

Captured-frame television: Television that is either *slow-scan* or *freeze-frame* variety.

Codec: A contraction of "coder-decoder." It is a device which converts an analog signal into digital form for transmission, and converts it back again at its destination. An important aspect of the encoding, especially for video conferencing, is the removal of redundant information. The encoding may also involve the combination of different signals (e.g., video, audio and graphics) which will be separated again in the decoding process.

Computer conference: A *teleconference* in which participants communicate in a text mode, using keyboards to enter messages. Messages are processed (routed, stored, etc.) by a computer. Transmission is generally via a *value-added network*. Although messages may be exchanged in real time, the majority are usually stored until intended recipients next log in.

Conference telephone: A device including a loudspeaker and one or more microphones, which, when connected to the telephone network, allows a group of users to hear and be heard by others in a teleconference. There are many kinds of conference telephones: some are portable; some employ "push-to-talk" microphones; some employ voice-switching to deaden the loudspeaker when the microphone is active.

Decibel: A unit of a logarithmic scale used for measuring the strength of a signal relative to some reference level. (The "loudness" of sound is often measured in decibels.) Since the number of decibels is 10 times the logarithm (to the base 10) of the ratio of signal strength to reference level, an increase of 10 indicates a tenfold increase in strength, an increase of 20 indicates a hundredfold increase, and so on.

Digital signal: A signal encoded as a series of discrete numbers.

Digital Termination System (DTS): A new transmission system which provides digital connections via microwave facilities from users' premises to shared earth stations. The system is intended to overcome the expense of *local ends* for users who do not have an on-premise earth station for satellite communications.

Duplex transmission: Transmission in which signals can flow in both directions at the same time.

Earth station: The antenna ("dish") and associated equipment used for transmitting signals to and receiving signals from a communications satellite. Some stations transmit and receive; others only receive.

Echo: A wave reflected with sufficient energy and delay to be perceived as distinct from the directly transmitted signal.

Electronic blackboard: A device for converting the pressure of chalk writing or drawing on a board into a signal for transmission over a telephone circuit and for converting the signal into an equivalent image on a monitor at the destination. (The term usually refers to the Bell System's Gemini™ 100 electronic blackboard.)

Enhanced audio conference: A teleconference employing both audio and some electronically controlled or transmitted graphics. Graphics may be provided by remotely controlled slide or microfiche projectors, *slow scan* or *freeze-frame* television, or an *electronic blackboard* or table.

Field: In television, the information about an image conveyed in the alternate (odd or even) scan lines. In the standard interlaced scanning system, two fields (one of odd, the other of even scan lines) are required for each frame.

Frame: In television, the total transmitted information in a scanned image. One frame consists of two interlaced *fields*.

Freeze-frame television: *Slow-scan television* with the added feature that the image to be transmitted is "frozen" in a local memory prior to transmission. With slow-scan television, movement of the image during the transmission (which may take 30 seconds or longer) results in a blur. This problem is avoided with freeze-frame television.

Gain: In transmission, the increase in the power of a signal between two points.

Half-duplex transmission: Transmission in which signals can flow in either direction, but in only one direction at a time.

Hertz: A unit of frequency equal to one cycle per second (CPS).

Local area network: A private transmission network generally interconnecting offices within a building or a set of nearby buildings and usually designed to convey different kinds of traffic: e.g., voice, data, facsimile and video.

Local ends: Transmission links between customers' premises and trunk circuits, e.g., the connection to an off-premise *earth station.*

Loss: In transmission, the decrease in the power of a signal due to resistance or impedance as it passes through a circuit or equipment.

Meet-me bridge: See *bridge.*

Multiplexer: A device which combines multiple signals for transmission via a common channel.

Noise: Unintended signal introduced by circuit components or natural disturbances.

Operator-assisted bridge: See *bridge.*

Packet-switched network: A network in which messages are divided into packets to which headers are added. The header includes information as to the destination and information necessary for the reconstruction of the message at its destination. The packets are transmitted separately from node to node through the network. At each node a computer determines the best onward routing. The separate packets may travel different routes from origin to destination. By increasing complexity, packet-switching allows for an improved balance between network utilization and the avoidance of congestion.

Pixel: A contraction of "picture element." At any instant, a video screen is a rectangular array of pixels, each of which has a particular level of illumination (and color, in the case of color television). A 525-line television image comprises approximately 330,000 pixels.

Point-to-multipoint video conference: An asymmetric video conference in which all sites receive television images, but only one site can transmit them. The audio component may be bidirectional or unidirectional. Television transmission is usually via a communications satellite; if the audio is bidirectional, it is likely to be transmitted via the public telephone network.

Point-to-point video conference: A video conference between two sites, each of which can transmit and receive both television and audio.

Port: A point of connection of a line to a bridge. The number of ports on a bridge is the maximum number of lines it can interconnect.

Private branch exchange (PBX): A private switching system for interconnecting a customer's internal telephone lines ("extensions") with one another and with the public telephone network.

Resolution: The quality of a television image which allows the observer to distinguish detail. Resolution increases as the number of *pixels,* hence number of lines, increases. In North America, 525 lines is the standard. There are, however, systems which display 1000 or more lines; they are termed "high resolution" systems.

Signal: Information which has been encoded, usually in electromagnetic form, for the purpose of transmission.

Simplex transmission: Transmission in which signals can flow in only one direction.

Slow-scan television: A means of transmitting a still video image via a channel of a lesser bandwidth than is required for regular motion video. Typically this is a 3-kilohertz channel (i.e., a regular telephone connection). The information comprising the image is "trickled" down the channel and reassembled as a still frame at the destination. Resolution may be traded off against transmission time, but 30 seconds is typical for a black-and-white system; color images require more time.

Teleconference: The use of a telecommunications system for communicating with three or more people at two or more locations. The system used must allow real time communications though it may be used in an asynchronous (not real time) mode, too (e.g., computer conferencing).

Value-added network (VAN): A network operated by a private company which "adds value" to basic telecommunications services leased from common carriers, and resells the enhanced services to end users. Examples are the *packet-switched networks,* Telenet and Tymnet.

Video conference: A *teleconference* in which full motion video is transmitted, as well as voice and maybe graphics. The video signal can be one-way (from one point to many points) or two-way (simultaneously connecting two or occasionally more than two sites). The term is also used by some to include teleconferences employing audio plus *freeze-frame television.* Used interchangeably with "video teleconference."

Bibliography

BOOKS AND REPORTS

Center for Interactive Programs, *Teleconferencing and Electronic Communications: Applications, Technologies and Human Factors.* Madison, WI: University of Wisconsin-Extension, 1982.

Center for Interactive Programs, *The 1982 Teleconferencing Directory.* Madison, WI: University of Wisconsin-Extension, 1982.

Degnan, Kim E.; Pottle, Jack T. and Bortz, Paul I., *Teleconferencing: Industry Overview; Economic Parameters; Public Broadcasting Potential.* Report to the Corporation for Public Broadcasting. Denver, CO: Browne, Bortz and Coddington, 1981.

Federal Trade Commission, *Media Policy Session: Technology and Legal Change.* Staff report for policy session. Washington, DC, December 1979.

Hansell, Kathleen J.; Green, David and Erbring, Lutz, *Videoconferencing in American Business: Perceptions of Benefit by Users of Intra-Company Systems.* McLean, VA: Satellite Business Systems, May 1982.

Henderson, Madeline M. and McNaughton, Marcia J., eds., *Electronic Communication: Technology and Impacts.* AAAS Selected Symposium 52. Boulder, CO: Westview Press for the American Association for the Advancement of Science, 1980.

Johansen, Robert; Vallee, Jacques and Spangler, Kathleen, *Electronic Meetings: Technical Alternatives and Social Choices.* Reading, MA: Addison-Wesley Publishing Co., 1982.

Martin, James, *Future Developments in Telecommunications*. Englewood Cliffs, NJ: Prentice-Hall, 1977.

———, *The Wired Society*, Englewood Cliffs, NJ: Prentice-Hall, 1978.

Public Service Satellite Consortium, *Teleguide: A Handbook for Video Teleconference Planners*. Washington, DC: Public Service Satellite Consortium, 1981.

Satellite Business Systems, *Prelude: Communications of the 80's*. McLean, VA: Satellite Business Systems, 1978.

Videomeetings, *Down-to-Earth Satellite Meetings: A New Alternative in Corporate Communications*. Dallas, TX: Communications Technology Corp., 1981.

SELECTED ARTICLES

Biomedical Communications. Three-part series, "Video Teleconferencing Techniques." March/April 1982, May/June 1982, July/August 1982.

Charles, Jeff, "Approaches to Teleconferencing: Towards a General Model Justification." *Telecommunications Policy,* December 1981.

Costello, Marjorie, "Planning Your Videoconference." *Videography,* May 1981.

Cross, Thomas B. and Thomann, Daniel C., "Using Computer Conferencing in Telecommunications Management." *Business Communications Review,* May/June 1982.

Cutting, Bill, "Anatomy of a Videoconference." *Marketing Communications,* October 1981.

E-ITV, July 1981 issue. Seven articles on teleconferencing. Authors include Kathryn Bradford, Derek Mustow, Julie Meier Wright, Willard and Cinda Thomas and Don Wylie.

E-ITV, August 1982 issue. Five articles on teleconferencing. Authors include Eileen Bodie, Willard and Cinda Thomas, Robert M. Patterson, and Marcia Baird and Mavis Monson.

Elton, Martin C.J., "The Practice of Teleconferencing," in *Telecommunications in the United States: Trends and Policies.* Edited by Leonard Lewin. Dedham, MA: Artech House, Inc., 1981.

Glazer, Sarah, "Satellite Communications: Are These Services in the Cards for Your Company?" *Telematics,* July 1981.

Green, David and Hansell, Kathleen J., "Teleconferencing: A New Communications Tool." *Business Communications Review,* March/April 1981.

Hallam, Rob, "Ford Had a Better Idea — A Satellite Press Conference." *E-ITV,* January 1982.

Information Industry Review. Two-part series, "The Teleconferencing Trend: Keeping Your Distance May Pay Off." January 1982, February 1982.

Jenkins, Thomas M., "What It Takes to Teleconference Successfully." *Administrative Management,* October 1982.

Johansen, Robert and DeGrasse, Robert, "Computer-Based Teleconferencing: Effects on Working Patterns." *Journal of Communication,* Summer 1979.

Johansen, Robert; Hansell, Kathleen J. and Green, David, "Growth in Teleconferencing — Looking Beyond the Rhetoric of Readiness." *Telecommunications Policy,* December 1981.

Johnson, James W., "A Case Study: The Healthy Pulse of Medical Video Teleconferencing." *Satellite Communications,* May 1981.

Josephson, Sanford, "Dramatic Growth Looms for Videoconferencing." *Television/Radio Age,* June 15, 1981.

Paulsen, Gregory W., "Corporate Teleconferencing: An Investment for the Future." *The Office,* November 1980.

Pensinger, Glen, "How to Build Profits with Teleconferencing." *Broadcast Communications,* September 1982.

Pereyra, Susan G., *Meetings by Teleconference.* Danbury, CT: Connex International, Inc., 1980. Pamphlet.

Perham, John, "Business' New Communications Tool." *Dun's Review,* February 1981.

Pomerantz, David, "The Ins and Outs of Teleconferencing." *Today's Office,* July 1982.

Schubin, Mark, "Do You Need a Teleconference?" *Videography,* May 1981.

Showker, Kay, "Do Teleconferencing and Travel Mix?" *ASTA Travel News,* July 15, 1982.

Smith, Judson, "Teleconferencing: What It Is, What It *Will* Be and How It Can Help Trainers." *Training/HRD,* October 1979.

Telephony, July 26, 1982 issue. Four articles on teleconferencing, by Gordon B. Thompson, Elliot M. Gold, Daryl Braun, and Kathleen J. Hansell, David L. Green and Lutz Erbring.

Telephony, August 2, 1982 issue. Seven articles on teleconferencing, by Patrick S. Portway, Charles A. Pleasance, Elliot M. Gold, Thomas B. Cross, Ed Morris, and Kathleen J. Hansell, David L. Green and Lutz Erbring.

Thomas, Willard and Cinda, "Estimating a Teleconferencing Budget." *E-ITV,* August 1981.

Weber, David, "Making Waves in Teleconferencing." *Venture,* May 1982.

Woolf, Mary M., "A Telemeeting That Missed the Mark." *E-ITV,* November 1981.

JOURNALS, MAGAZINES AND NEWSLETTERS

Administrative Management, Geyer-McAllister Publications, New York, NY. (monthly)

Audio-Visual Communications, United Business Publications, Inc., New York, NY. (monthly)

Broadcasting, Broadcasting Publications, Inc., Washington, DC. (weekly)

Business Communications Review, BCR Enterprises, Inc., Hinsdale, IL. (bi-monthly)

Communications News, Harcourt Brace Jovanovich Publications, Geneva, IL. (monthly)

Computerworld, CW Communications, Inc., Newton, MA. (weekly)

Data Communications, McGraw Hill, New York, NY. (bi-monthly)

Datamation, Technical Publishing, New York, NY. (monthly)

Educational and Industrial Television (E-ITV), C.S. Tepfer Publishing Company, Inc., Danbury, CT. (monthly)

Information Industry Review, International Information/Word Processing Association, Willow Grove, PA. (monthly)

InterMedia, International Institute of Communications, London, UK. (bi-monthly)

Journal of Communication, Annenberg School Press, Philadelphia, PA. (quarterly)

The Office, Office Publications, Inc., Stamford, CT. (monthly)

PSSC Newsletter, Public Service Satellite Consortium, Washington, DC. (monthly)

Satellite Communications, Cardiff Publishing Co., Denver, CO. (monthly)

Satellite News, Phillips Publishing, Inc., Bethesda, MD. (bi-weekly)

Successful Meetings, Bill Communications, Inc., Philadelphia, PA. (monthly)

Telcoms, Center for Interactive Programs, University of Wisconsin-Extension, Madison, WI. (eight per year)

Telecommunications, Horizon House, Dedham, MA. (monthly)

Telecommunications Policy, Butterworth Scientific Ltd., Surrey, UK. (quarterly)

Telematics, Information Gatekeepers, Inc., Brookline, MA. (bi-monthly)

Telephony, Telephony Publishing Corp., Chicago, IL. (weekly)

The TeleSpan Newsletter, TeleSpan, Altadena, CA. (monthly)

Television/Radio Age, Television Editorial Corp., Easton, PA. (weekly)

Video Systems, Intertec Publishing Corp., Overland Park, KS. (monthly)

Video User, Knowledge Industry Publications, Inc., White Plains, NY. (monthly)

Videography, United Business Publications, Inc., New York, NY. (monthly)

Index

ABOUT THE AUTHORS

Ellen A. Lazer is senior editor of the Communications Library and Video Bookshelf series of books and monographs for Knowledge Industry Publications, Inc. Previously, she was an editor with Praeger Publishers. Ms. Lazer is consulting editor of *Video in Health,* and has contributed chapters to other books and studies. In addition, she has written reviews and articles for various publications. Ms. Lazer chaired the special teleconferencing seminar of *Video Expo,* which was held in New York in fall 1982.

Dr. Martin C.J. Elton is professor of communications at the Tisch School of the Arts, New York University, and is founding chairman of the NYU Interactive Telecommunications Program. From 1972 to 1976 he was director of the Communications Studies Group, University College, London, which was the world's leading noncorporate research center on teleconferencing. Dr. Elton is editor of *Evaluating New Telecommunications Services* and a contributing author of *Telecommunications in the United States: Trends and Policies,* and has written numerous articles and working papers on teleconferencing. He has led research studies on teleconferencing for many organizations, including AT&T, British Telecom and the Canadian Department of Commerce.

James W. Johnson is founder and president of TeleConcepts in Communications,Inc., a New York firm that specializes in the production of private television events and interactive closed circuit conferences. Prior to founding TeleConcepts in Communications in 1975, Mr. Johnson held executive positions with a number of video firms, before which he spent 14 years with CBS, Inc. in New York. Since 1975, Mr. Johnson has produced national and international teleconferences for such companies as J.C. Penney, IBM and Bristol Laboratories, and produced the first live international interactive video fashion show. He is the author of numerous articles on video production and closed circuit video conferencing, and is active as a speaker and lecturer.

Al Bond is manager of the Texas Instruments Media Center, Dallas, TX. Previously, he was a news media producer at the Manned Spacecraft Center, Houston, TX, where he worked with TV and radio networks worldwide to provide coverage of the Gemini and Apollo space programs. He is a contributing author of *The Video Handbook* (London) and has published articles in a number of professional journals. Active in video conferencing since 1976, and in satellite conferencing since 1980, Mr. Bond is presently chairman of the board of directors and director of international affairs for the International Television Association.

A. David Boomstein is national project coordinator—teleconferencing, for the corporate telecommunications department of Citibank NA. Previously, he worked in the telecommunications, communications media and cable television fields, and is a graduate of the first class of New York University's graduate program in interactive telecommunications. Mr. Boomstein is currently president of the International Teleconferencing Association.

Eugene Marlow is president of Media Enterprises, a communications firm that specializes in the use of video for information and entertainment purposes, and conducts studies on the potential of cable television for business communications. Prior to forming Media Enterprises, Mr. Marlow was a consultant to and head of communications functions for Union Carbide Corp., Prudential Insurance Co. and Citibank. As head of video communications at Union Carbide, he expanded a worldwide video network and created a large-scale media department which garnered more than 40 awards for excellence, including "Department of the Year" honors from the Information Film Producers Association. Mr. Marlow is author of *Managing the Corporate Media Center* and a contributing author of *The Video Handbook* (London).

Robert D. Rathbun is editor of the International Council of Shopping Centers publications. Previously, he was managing editor of the trade publications *Video User* and *Shooting Commercials*. Earlier he was station manager for Cable Channel Nine in Lexington, VA, and has worked as a newspaper reporter and editor for the Berkshire Courier in Great Barrington, MA, and as a broadcast reporter for WDBJ-TV in Roanoke, VA. Mr. Rathbun is a contributing author of *Video in Health,* and has written for a number of publications.

Bonnie Siverd is a freelance writer specializing in business and finance. Previously, she was a staff editor for *Business Week* magazine. Prior to joining *Business Week*, she wrote for the *Congressional Quarterly* and *Time* magazine. Ms. Siverd is author of the forthcoming book *Personal Finance for Women,* and is a contributing editor for *Working Woman* magazine, and also writes the magazine's "Money" column.

Other Titles Available from Knowledge Industry Publications

The Home Video and Cable Yearbook, 1982-83
263 pages   softcover $85.00

The Video Age: Television Technology and Applications in the 1980s
264 pages   hardcover $29.95

The Future of Videotext: Worldwide Prospects for
Home/Office Electronic Information Services
by Efrem Sigel, et al.
197 pages   hardcover $34.95

The Video Register, 1982-83
320 pages   softcover $47.50

Handbook of Interactive Video
edited by Steve and Beth Floyd
168 pages   hardcover $34.95

Video Discs: The Technology, the Applications and the Future
by Efrem Sigel, Mark Schubin, Paul F. Merrill, et al.
183 pages   hardcover $29.95

Office Automation: A Glossary and Guide
edited by Nancy MacLellan Edwards
275 pages   hardcover $59.50

The Word Processing Handbook: A Step-by-Step Guide
to Automating Your Office
by Katherine Aschner
193 pages   hardcover $32.95
                       softcover $22.95

Electronic Mail: A Revolution in Business Communications
by Stephen Connell and Ian A. Galbraith
141 pages   hardcover $32.95
                       softcover $22.95

Available from Knowledge Industry Publications, Inc., 701 Westchester Avenue, White Plains, NY 10604.